Roger Vergé's Cooking with Fruit

Roger Vergé's

Cooking with Fruit

Roger Vergé
with
Adeline Brousse

Translated by Molly Stevens

Photographs by Jean-Pierre Dieterlen
assisted by Danièle Schnapp

Harry N. Abrams, Inc., Publishers

Acknowledgments

Roger Vergé would like to thank all the members of the Flammarion team, as well as Adeline Brousse, Michel Duhamel, Serge Chollet, Daniel Desavie, Rémi Fünfrock, and Sylvie Charbit.

Danièle Schnapp would like to thank all the stores that provided the necessary props for the photographs: Baïkal (glassware by Rémi Muratore), Bon Marché, Boutique Dior, Boutique Scandinave, Christofle (Christian Lacroix dishes), Eric Dubois, Espace Buffon, Simone Ezagury (glasses and decor), Lunéville (pottery), Fuschia, Habitat, L'Entrepôt, the Loft, L'Orjol, Noces de Cana, Porto Santo, Primrose Bordier for the Jaquard Français, the Puceron Chineur, Siècle. The painted backgrounds were provided by Célia Eynard.

Flammarion would like to express gratitude to Marc Walter, Margherita Mariano, Murielle Vaux, Nathalie Bailleux, and Anne-Laure Mojaïsky for their assistance.

Art Director: Marc Walter
Editor, English-language edition: Elisa Urbanelli
Designer, English-language edition: Carol Robson

Library of Congress Cataloging-in-Publication Data

Vergé, Roger, 1930–
      [Fruits de mon moulin. English]
      Roger Vergé's cooking with fruit / Roger Vergé with Adeline Brousse ;
translated by Molly Stevens ; photographs by Jean-Pierre Dieterlen.
          p.   cm.
      Includes index.
      ISBN 0–8109–3931–2 (hardcover)
      1. Cookery (Fruit) 2. Fruit—France. 3. Cookery, French.
I. Brousse, Adeline. II. Title.
TX811.V4613   1998
641.6'4—dc21   98–3417

Printed and bound in Italy

Harry N. Abrams, Inc.
100 Fifth Avenue
New York, N.Y. 10011
www.abramsbooks.com

Contents

Fruits of the Market   7

Fruit Salads and Compotes   37

Tarts and Tartlets   67

Prepared Desserts and Baked Goods   93

Coulis, Fruit Drinks, and Sorbets   125

Preserves and Candy   145

Little Chefs in the Kitchen   175

Sweet and Savory Condiments   181

Conversion Tables   188

Index   190

FRUITS OF THE MARKET

A fruit plucked from its branch, still warm from the sun or sparkling with dew, both fragile and tender, is a wonderful thing to hold in our hands. Next, our sense of smell is aroused by the fruit's fresh aroma. In our mouths, the pulp, juice, and flavors are released to complete our delight. Thus, in just a few seconds, a fruit can stimulate and invigorate all of our senses. It should come as no surprise that biting into an apple is a metaphor for pleasure!

Fruit triggers an inner desire and evokes a taste for sugar that, for me, is associated with childhood. Many years and a career-as-chef later, I am still filled with nostalgia for the sugar on bread that I would munch as a child. Even then I was striving for the right proportion of sugar to bread, seeking the perfect balance between crunch and soft, between the intensity of the sucrose and the silkiness of the starch. My current desserts continue to reflect this perpetual search for harmony—between the sweetness of sugar and a fruit's sour note, between the juicy pulp soaked in syrup and the crispiness of a pastry crust.

Ripened fruits are filled with vitamins and minerals, and thus offer a health benefit. Most fruits contain a large amount of pro-vitamin A, which makes skin supple and accelerates tanning. The darker the pulp of an apricot, peach, melon, plum, or mango, the more vitamins it contains. Many fruits also provide vitamin C; it might come as a surprise to learn that strawberries have as much vitamin C as oranges, while black currants have four times as much. Kiwi not only contains vitamin C, but also magnesium and calcium, as do raspberries, which are also full of iron. Each fruit is an original source of various nutriments; sample them all, and sample the pleasures.

There is therefore no good excuse for resisting a fruit dessert—not even lack of time to make one. Fruits such as raspberries and apricots, which are as good plain as cooked, are often placed whole, skin and all, in tarts, salads, or side dishes. Sometimes peels and stems can be removed quickly, as is the case with bananas and strawberries. Fruit can be used easily in many quick dishes. Nor can you plead fear of failure: the art of making dessert is not so difficult, and the recipes in this book are written like cooking lessons, placing success within a beginner's reach.

Picking fruit in the orchard has always made me feel peaceful. When I was a child in Commentry, we went out every day during the summer to gather

Flavored Sugars

For a change of pace, play with adding all kinds of flavors to your sugar. Flavored sugars are wonderful simply sprinkled on desserts. Tightly sealed in jars, flavored sugars will retain their taste for many months.

Vanilla. *It is as easy as keeping 2 or 3 vanilla beans, split lengthwise, in 2 pounds of sugar, in a closed jar, for at least three weeks. The sugar will absorb the beans' abundant taste. Sprinkle over fruit salads, compotes, jams (pear and papaya, for example), and tarts.*

Star aniseed *(star anise). Prepare like the vanilla sugar, but bury 2 star anise in the jar.*

Cinnamon. *Prepare like the vanilla sugar, but with 2 sticks of cinnamon.*

Lavender. *Prepare like the vanilla, replacing the beans with about twenty sprigs of very fresh lavender. Let the jar sit for at least two weeks.*

Orange, lemon, or lime. *Same steps as above, burying the fresh zest of 2 or 3 oranges, lemons, or limes in 2 pounds of sugar. Let rest for at least two weeks.*

perfectly ripened fruit, once the first to fall to the ground urgently told us that the harvest had begun.

So that we would not have fruit that was wet or too warm, we would always go gathering either in the morning, our feet in the dew, or in the evening, under the golden cast of the setting sun. We would not go if it was too wet outside, when fruit runs the danger of rotting quickly. Rather, we waited until the fruit on the tree was dry. We carried several baskets when picking fruit and sorted the fruit according to quality. The healthy specimens, which included apples, pears, and quinces, were either laid out on fruit racks in the basement to become our winter provisions, or used right away in tarts and fruit salads. The wormy pickings, or those that were slightly spotted, would become compotes, preserves, and syrups.

We learned how to pick fruit properly, making sure not to hurt or damage the branch that held it. Otherwise, the branches would not be able to bear fruit the following year. The trick is to not pull on the branch. The stem has to detach itself easily from the branch when slightly twisted. On the other hand, you can tell that red fruits, such as wild strawberries and raspberries, are overly ripe if they fall by themselves.

Today we are more likely to gather fruit at the market. To do so successfully, you need to know how to compare quality among all the varieties of fruit that sit side by side on the stands. When it comes time to choose, remember that the biggest fruits are not necessarily the best. When possible, try to look at the skin and smell it. If the vendor has cut open one of his fruits, inspect it, comparing the outside of the sample with the others in the group to make sure they are the same.

Watch out for fruit that has been "ripened" through sorting; pinching and poking fruits to check their ripeness will bruise them. It is often better to use smell as a guide, assessing the fruit according to the potency of its fragrance. And do not store your purchases in a closed plastic bag, as condensation is good grounds for rotting. Keep the bag open or use brown paper bags that let the fruit breathe.

Spring Fruits

Strawberries

In France, light, flavorful strawberries always represent the first fruit of the season, welcoming spring in April and lasting until October. As a child, I would not even let their skin become shiny before picking them. My little hand would steal berries off the vine in the early morning, when the fruit was still pink, and I would relish their tartness.

For cooking, I choose from among the strawberries that are available at the time I want them. There are as many as twenty different varieties that appear at French markets one after the other, ensuring that the stands are never empty. During the winter, they come from South Africa or Israel. Round or cone-shaped, a rather bright red, these strawberries are all refreshing, not too sweet, and, in general, nicely scented. Their flavor is more or less accentuated depending on their maturity and on the weather.

Growers pick strawberries once ripe (they will not ripen further after picking) and ship them immediately to retailers. The berries will go bad in just a few days. Avoid touching them and, instead, look at them to make your selections. Choose those with very green and firm calyxes. Avoid those that are pale near the stem or that have pink or brown spots. Bring the strawberries close to your nose and choose the ones with the fullest aroma.

Wines to Drink with Fruit

It seems natural to combine wine, a pure reflection of grapes, with other fruits to create a perfect union of scent and flavor. Generally speaking, fortified or syrupy wines, of which there are many kinds, are good accompaniments for all fruit desserts.

• Sweet muscat wines that smell like fresh grapes: Rivesaltes and Beaumes-de-Venise.
• Fortified wines, whose aroma enhances spices, red fruit, dried fruit, and/or fruit preserves: Banyuls, Rivesaltes, Maury, and Rasteau. Similar to port.
• Wines made from raisins, which exude a strong aroma of dried fruit and are often very sweet: Italian Vino Santo, Vin de Paille from the French Jura, and Vin de Paille de Gérard Chave en Hermitage from the Rhône Valley.
• Syrupy wines from Bordeaux and southwest France, which smell of ripe fruit, flowers, and honey: Sauternes, Montbazillac, and Jurançon . . . or sweet white wines from the Loire, which smell like apple and quince: Bonnezeaux, Montlouis, and Vouvray.
• Tokay-Pinot Rosé and Gewurztraminer from Alsace, known as "late harvests," "noble rotted sweet wine," or Eiswein ("ice wine"), offer the perfect harmony of sweet and sour, echoing the balance of fruit. Note the vintage Christine des Domaines Schlumberger, Léon Beyer's Gewurztraminer Eiswein, and Jean Hugel's late Riesling. Rieslings sometimes do wonderfully well in California's Napa Valley, like those from Robert Mondavi.
• Hungarian Grand Tokays, with their handsome orangelike color and their scent of dried apricots.

Certain wines go especially well with a particular fruit:
• With apricots: Hungarian Tokays, Banyuls, White Banyuls, Rivesaltes, Maury, Beaumes-de-Venise, Jurançon, Pacherenc de Vic-Bilh, and Montlouis.
• With apricots, nectarines, and peaches: Condrieu and Château-Grillet, and dry white wines from the Côtes du Rhône.
• With melons and strawberries: Banyuls, Rivesaltes, and ports.
• With peaches: Jurançon and muscat wines.
• With mangoes and litchis: Sauternes and Barsac.
• With raspberries: A Saint-Estèphe Médoc.
• Sipping an old Médoc with strawberries is a Bordeaux tradition.

Do not keep strawberries in the refrigerator. They should be eaten the same day they are purchased. Rinse them briefly in a colander under the faucet so that they do not sit in water. Dry them with a clean cloth and then remove their stems. You will find that the berries have more flavor when served at room temperature. When eaten immediately, they provide an abundance of vitamins.

Strawberries do not freeze well. They barely survive in good enough condition to be mixed into a coulis. To preserve them, make jams, compotes (to be frozen), or syrups.

If during a walk through the fields in June or July, or in the mountains during August and September, you are lucky to come across the small variety known as wild strawberries, make sure to taste them. They are so small you will be tempted to eat five or six at a time. Their smell and taste, strong and almost tart, exemplify why I so often prefer small fruit to big. Sometimes wild strawberries can be found in little baskets at the market.

Cherries

Cherries, the first tree fruit to be picked in the spring, are an irresistible temptation. Their thin, tight skin, smooth and shiny, ranges in color from pale yellow to dark red, depending on the variety. They can be eaten straight from the branch. Their brief season makes them all the more

Flowers and Fruit

When I create a dessert, I look for the best blend of texture, taste, scent, and color, in order to create a thrilling bouquet. Often, flowers help me to perfect an elegant composition. Their flavor can strengthen or complement the fruit, and their color and beauty add a vibrant or delicate touch.

Roses

You will find rose water in gourmet food stores. Add this delicate flavor to classic wild strawberry, raspberry, or mixed red-fruit salads, as well as to peach-based desserts and white nectarines. In Tunisia, a country of subtle flavors, rose water is mixed with pomegranate. This unique combination is surprising and beautiful. Open up a ripe pomegranate and extract the seeds soaked in pulp. Add grenadine syrup and a small amount of rose water to these juice-filled rubies. Serve very chilled in glass bowls. First, dip the edges of the bowls in thinned grenadine syrup and then in sugar; the bowls should now have a delicate, pink-sand collar. Finally, stand the bowls on a bed of freshly picked rose petals.

Violets

Good violet water can be found at gourmet shops, along with rose water. Not well known and uncommon, it goes wonderfully well with buttery, thinly sliced, ripe pears and a touch of lemon. Fill small bowls with this very simple salad and decorate with fresh violets in the spring, or crystallized ones in the off-season (they can be bought in candy stores). To enrich the marriage of colors, garnish with a few violet leaves: the small leaves that can be gathered in the hinterland of Provence are edible; those from elsewhere should be removed from your plate and left uneaten.

Borage

Admired for their color, these small blue flowers that bloom in springtime are often used in northern Italy for decorating fruit tarts. They create a nice effect when arranged around a strawberry tart, for example. The easiest way to acquire borage is to plant it in your garden or out on your patio; you will find the plants at your nursery.

Acacia Flowers

Powdered with confectioners' sugar and placed in clusters on prepared desserts or tarts, this flower brings a touch of purity, elegance, and grace. I have used them to make delicious fritters.

Jasmine

Fresh white jasmine flowers will wonderfully complement an orange salad's bouquet or any fruit salad with oranges, melon, or papaya. Arrange the flowers at the last moment—no more than 15 minutes before serving—because they tend to turn brown. When garnishing salads with jasmine, place thin branches with flowers around the dish or bowl, like a necklace. But be warned: jasmine pleases the eyes and nose but not the stomach. Although they are edible, they usually are not eaten. If you do not plan to eat them, buy them from the florist. Otherwise, pick them from places where they have not been treated with chemicals.

Lavender

Lavender is beautiful lightly scattered on compotes, in which case I prefer to use honey instead of sugar. Do not use the flowers found in sachets or potpourri bags; their aroma has been intensified artificially, making them inedible. The ideal is to pick lavender yourself from the garden or in the wild. Dry it for use all winter long. Touching these flowers might be unpleasant; if I use them with honey, I pulverize them into powder. Otherwise, I crush them with sugar (1 part flowers to 12 parts sugar).

Orange Blossoms

Orange blossoms are wonderful placed alongside fruit, thanks to their clean, pure scent. I add them to dishes at the very last moment. Orange blossom water, sold in pretty bottles at gourmet food stores, is also delightful. I sometimes add it to the cream that I use in prepared desserts. If you can find the fresh blossoms, scatter them on fruit salads.

desirable; they arrive in mid-May, make a short round at the market, and are gone by July. In Spain, their production is prolonged until the beginning of August. Cherries from Chile are imported into France for two or three weeks during the winter, surprising us around the holidays.

If you gather cherries from the tree, wait until they have attained their most intense color; it is then that they will have produced their capacity of sugar, vitamin C, and provitamin A. Hurry, or else birds will help themselves before you do. Pick cherries with their stems and lay them in flat baskets, being careful not to stack them too high so that they will not get crushed. If purchasing cherries at the market, keep in mind that cherries are fragile and will not last more than

two or three days after they have been picked. A bright green, straight, and supple stem is the sign of a fresh cherry. Cherries do not do well when stacked up or stored in the refrigerator, where condensation will make them rot. For the same reason, they should not be washed too much in advance of using them. Buy them regularly, as needed, in small quantities.

Cherries can handle freezing for up to ten months, as long as you wash and dry them well, and remove their stems beforehand. Definitely do not pit them. Once winter comes, sprinkle cherries in flans, clafoutis, and tarts while they are still frozen, so as not to lose a single drop of juice before or during cooking.

You can no longer buy the wild cherries that dazzled me as a child—not even their equivalent. In central, eastern, and southern France, foragers still hunt for small cherries or morellos growing wild. Do not hesitate to pick these delightful fruit, which are sometimes acidic and bitter, and sometimes very sweet. Wild cherries can be used interchangeably with cultivated varieties in my recipes, but never try to pit them. Their accentuated scent and flavor will add a distinctive character to your dish.

Sweet cherries are all related to the wild cerise. Gean cherries have a tender flesh which make them especially right for preserves, compotes, and blended drinks. Bigarreau cherries, with their firm flesh and solid skin, can be eaten plain or thrown into recipes that call for whole cherries. A range of different varieties can be found in France at different times in the season: Burlat, Stark Hardy, Van, Summit, Reverchon, Coeur de Pigeon, Napoleon, Sunburst, Marmotte, Hedelfingen, and Duroni.

Morellos, or sour cherries, are less common. Practically the only morellos on the market are frozen, pitted ones, which I think are very good, with their rather firm, tasty flesh. Red morellos and the related amarelles (among them, the Montmorency cherry) can be cooked with brandy or used in clafoutis and other baked desserts. English cherries, however, are kept for canning, jellies, and liqueurs.

Rhubarb

Rhubarb's fat and curly buds flower along the ground at the end of winter. With only a few rays of sun they unfold wonderfully in only a few days. The full-blown plant will have very large leaves with wavy edges, supported by robust stalks. The stalk is the part we eat. It is a good idea to plant rhubarb in your garden, as it grows easily and is decorative.

An early fruit, rhubarb does not like excessive heat. We find it at the market between May and July, returning shortly in September and October. When it is good and ripe, its plump and firm stalk turns a purplish red with green accents. This is when I prefer it: the mature fruit is more flavorful, juicier, and more tart than the early fruit. The leaf's vigor will tell you how fresh the plant is, as will the base of the stalk. The stalk should be brittle, allowing some juice to ooze out.

To peel, grab the thin transparent skin at the base end and pull toward the top. If the leaves have been removed but the stalks are still unpeeled you can keep them for two or three days in the vegetable bin of your refrigerator; longer than that, they will dry up and soften. However, rhubarb freezes very well. Simply wash it, chop it into pieces, dry it completely, and wrap it in plastic.

If you wish to reduce its bitterness, combine rhubarb with apple or scald it for two minutes before using it in a prepared dessert, a tart, or a compote. I also use rhubarb in preserves, syrups, ice cream, and drinks (like the Italian *rabarbaro*). I love to blend rhubarb with strawberries to create a wonderfully tangy taste. A rhubarb compote is delicious with fish.

Apricots

I should tell you right away, apricots are my favorite fruit. I love their silky skin, golden like the sun, especially when it is tinged with red, like a plump cheek. I love the juicy and tender flesh between my teeth and the aroma of the fresh fruit, as well as its kernel and nectar, which tastes both sweet and a bit sour. I even like to suck on the smooth, practically flat pit. I crack it in search of the kernel, which I then peel and use to flavor my recipes.

I only use an apricot when it is good and ripe, making it one of the easiest foods to eat. If not ripened, an apricot will have no taste. When overripe, it becomes mealy and bruised, and sometimes has a horrible smell.

Unless you are able to pick apricots from the tree at just the right moment, it is not easy to find this delicious fruit at its ideal stage. Being delicate, apricots are harvested a little early so that they can better survive shipment. Once separated from the branch, however, they will remain hard and will always lack sugar. You will need some patience to find the ripest ones, making sure they are not blackened with spots or have not softened with time. Because they turn their final color early, you must rely on their supple but not-too-soft pulp and their full scent to judge ripeness.

Apricots must be eaten immediately. If you wait, they will quickly dry up, wrinkle, and become mushy. Do not stack them on other produce, and avoid putting them in the refrigerator, where they will dry out. If you must, seal them first in an airtight container, but not for more than two or three days. Apricots freeze well: cut them in half, remove the pit, and freeze the halves lined up flat. Once hardened, group them in a plastic bag.

When I was a child, the apricot season lasted only a month and a half. It is longer now, stretching from May until almost the end of June, thanks to the different, often new varieties that arrive one after the other. In France, we get apricots first from Tunisia, Greece, and Spain, then from Languedoc-Roussillon, then from Provence, and finally from the Rhône Valley. Apricots can also be found to the north of the Loire, often at a later date.

Sample the dark orange *lambertin* that arrives before the big, copper, almost red variety, with its firm, musky flesh. The huge, dense, and tart *jumbocot* is ideal for stuffing. The small red apricots from Roussillon are completely round and very flavorful. The large orange-colored variety from Provence is spotted with red; the big *bergeron* is recognizable by its half-red, half-orange coloring. Growers are researching bigger types, as well as later-growing varieties that will prolong the season.

Personally, I do not think bigger is necessarily better, and the apricots I prefer are not so easy to find. I have a soft spot for the rare, small *muscat*, which is longer in shape, with red spots,

and tastes like honey. If you happen to be in the south of France, you might find them by chance in a market. Don't pass them by! Whatever the variety, apricots can be used to make all kinds of treats: fruit salads, tarts, compotes, preserves, coulis, syrups, and even savory dishes—they are good with pork or veal.

Summer Fruit

Watermelon

This large, balloonlike melon, pale green and sometimes striped or marbled with yellow, depending on the variety, is related to the gourd in the cucurbit family. Its red flesh speckled with big black seeds is a wonderful source of refreshment on a hot summer day. In fact, watermelon is like a watery gourd; ninety-two percent of the fruit is vitamin- and mineral-filled water. Easy to ship, it is the perfect fruit for country lunches and summer picnics. Watermelons do not need to be refrigerated since, when still uncut, they keep their juice fresh naturally.

New kinds of watermelons offer certain advantages: for example, some are seedless (Queen of Hearts, Jupiter, Reina, Negra). Within the Golden Crown's oval rind, both smooth and striped with yellow, lies a handsome red interior with seeds lined up in a row so orderly that they can be removed with the single pass of a knife!

Between the end of April and the end of August, watermelons are often sold in pieces so that you can see their insides. Look for an intense and firm pink flesh. The light-colored strip between the flesh and rind should not be wider than a half inch. Do not buy slices that have been sitting in the open air; as soon as it is sliced, watermelon needs to be wrapped in plastic. If you are buying a whole melon, lift it up to check that it is suitably heavy.

Watermelons can be frozen like other melons (applying sugar is optional), a boon for winter fruit salads. I turn watermelon into a delicious preserve, combining it with a touch of bitter orange.

Peaches and Nectarines

A peach is one of the prettiest summer fruit, with its colors softened like a watercolor by its downy skin. Delicate and elegant, peaches range from the lightest pink to a practically black red, and from pale yellow to orange. Their firm, supple cheeks have inspired botanists to grant very feminine names to the different varieties of peaches; known as "nipples of Venus" under Louis XIV, today peaches are called Alexandra, Aline, Daisy, Melody, Primrose, and Tenderness. Including nectarines, there are more than four hundred kinds on the market! Only professionals know them well. Their varying maturity periods allow the season to stretch from June to the end of September, although each variety might only be available for a couple of weeks.

Gourmets distinguish peaches by color. I have an ongoing weakness for white peaches, which were among my favorite treats as a child. I liked to let the abundant juice run down my chin as it gushed from the glistening flesh. The peach melted so quickly in my mouth, I barely had time to chew it. Unfortunately, white peaches are fragile and cannot withstand shipping when ripe. In cooking, I use them to make compotes, drinks, and sorbets, and in recipes that call for whole peaches.

The firmer, yellow peaches are a little less juicy but easier to work with. Through the progress of natural selection they have become quite similar in taste to the white ones. I prefer yellow peaches for recipes that call for pretty slices, or for those that require fruit that does not fall apart while cooking.

The small vine peach, sanguine in color, is a cut above the rest. Not very sweet, its flesh, reminiscent of the color of wine dregs, has a sharp flavor, just like a good red wine. Is this coincidence or an incredible example of environmental influence? In France, many peach trees grow amid the vineyards of the Rhône Valley and on the hillsides of Lyons, and the fruit ripens at the same time as the first grape harvests! Because these small peaches are very decorative, I use them in my recipes for their color. Their solid taste balances out other, sweeter fruit.

Nectarines are a peach variety that appeared centuries ago, a result of a natural mutation process that caused them to lose their fuzziness. The white and yellow varieties share the same characteristics as peaches of the same color. Their difference lies in their flesh, which is generally firmer than that of peaches, making them easier to slice and facilitating handsome presentations. Nectarines do not lose their shape in the oven and make thicker preserves. Some varieties of nectarines are easier to pit than others.

Peaches are one of the most refreshing summer fruits. They are filled with water and are often not too sugary, even though they vary

Herbs and Fruit

Famous chefs, in search of whimsical tastes, use herbs such as thyme, rosemary, or basil in desserts.
I am not always wild about these experiments. However, I welcome certain herbs in my recipes,
those whose flavor, in my mind, truly complements that of the fruit.

Bay Leaf

A bay leaf is suited perfectly to a fruit syrup cooked with wine and featuring peaches, pears, or prunes. Combine bay leaf with the following spices: cinnamon, vanilla, and black pepper. Together they create a rich and complex effect.

Mint

I like mint's fresh taste in salads. Peppermint, in particular, enlivens certain exotic fruits, such as pineapple and papaya, adding both strength and finesse.

I like spearmint, but my favorite is sweet mint, the most common kind, which grows in the herb garden next to my restaurant, the Moulin.

Mint remains very flavorful when dried and therefore can be found all year round. If using dried mint you will need only half the amount.

Verbena

Verbena is a perfect accompaniment to apricots, as well as peaches and nectarines, especially with syrup. Its slight lemon flavor complements vanilla very well. Verbena leaves supply delicate decoration for fruit salads.

Fresh verbena cannot be replaced by dried verbena (verveine), which is used in teas. However, you can use verbena liqueur, the kind from Velay. Its flavor is so concentrated you need only 1 or 2 drops.

Tea Plant

The tea plant is not really an herb, but I use it like an herb. When combined with fruit, it forms a symphony of subtle tastes. I flavor compotes, salads, and cooked fruit syrups with bergamot tea (Earl Grey), jasmine tea, and vanilla tea.

greatly in flavor: some seem very sweet while others are much more tart. When I want to eat a peach out of hand, I simply wash it in water and avoid peeling; most of its vitamins are in the skin. However, peaches do need to be scalded briefly and then peeled if they are going to be used in prepared desserts.

Scientists have counted eighty elements constituting the peach's wonderful bouquet. They develop suddenly, right before the fruit reaches its maturity. Therefore, the fruit will have no scent unless it is perfectly ripe. If peaches are picked too early, the scented components will stop developing. There is no use in trying to ripen a peach at home! Your only recourse with an unripe peach is to poach it in syrup, which will give it the sugar it is missing. Pick peaches from the tree as soon as they are ripe. Two or three days later, they will already have begun to rot.

Do not touch the peaches at the market to see if their flesh is tender. Each pressing finger will ruin them. Nor should you rely on their color, as the skin often takes on its color very early. There is only one way to find ripened peaches: choose the ones with the most scent. To benefit even more from this aroma, store peaches at 65°F in a flat fruit bowl, and do not pile them up. You can also put them in the refrigerator (45–50°F) as long as they are returned to room temperature three to four hours before eating.

To freeze peaches, scald, peel, and cut them in half. Remove the pit and sprinkle the halves with sugar. Lay them flat and freeze them until they are hard. Then they can be grouped in a plastic bag. Allow them to defrost slowly at room temperature.

Raspberries

Although they have grown wild for hundreds of years, raspberries have been mass cultivated only since the early twentieth century. Many varieties have been eliminated as a result. For example, in the past you could often find golden yellow raspberries, such as those in my father's garden, among the other types. The yellow raspberry, although very good, has practically disappeared. We are therefore left to enjoy only the red variety. In France, patient growers cultivate them in the Val-de-Loire, the Rhône Valley, the Limousin, and the Île-de-France regions, and this does not include all the Sunday gardeners who nurture them in their backyards. Between the beginning of June and the end of September, each variety of these oval or cone-shaped fruit, varying in darkness, will provide a wonderfully scented flesh, sweet but also a touch sour. I am not as fond of the fruit raised in greenhouses, which appears as early as mid-April. Imported raspberries from Spain and Chile, available in the winter and spring, are generally acceptable. I avoid the American loganberry (Logan raspberry), a hybrid of blackberry and raspberry, which, although beautiful and very plump with dark skin, is unfortunately bland.

Be wary of very dark raspberries. When too ripe, the berries on the bottom of the container run the risk of being damaged during shipping or of becoming moldy. This fragile fruit does require some careful handling. As soon as you return from the market, empty the basket and spread the raspberries out on a plate. Eat them the same day. They can be kept for no more than thirty-six hours in the refrigerator, loosely stored in a sealed box. If crammed in, they will get moldy.

If cooking raspberries for jams or freezing them for later use, you need to act swiftly. The pectin contained naturally in the flesh quickly loses its ability to gel once the raspberries have been picked. Freezing will not damage raspberries, as long as they are not touching each other. Lay them out flat and freeze them until they are hard before putting them together in a bag. Their shape and flavor will be preserved if you defrost them slowly at room temperature and spread them out again. I keep any lost juice and cook it with sugar, then pour it over ice cream.

Red Currants and Black Currants

Even good and ripe, red currants are tart. Do not pick them as soon as they turn their vermilion red—they need another two weeks to produce sugar. It is a little more difficult to judge the ripeness of white and pink currants, which follow the same process and have close to the same flavor. Simply taste them before picking!

Red currants and black currants are part of the same botanical family. However, while red currants have a playful appearance, some people are surprised and put off by the inky skin and juicy insides of a black currant. Black currants are slightly sweet but tart, and a bit astringent because of the abundance of strongly flavored tannins. Therefore, make sure to pick them when ripe, right before the berries begin to fall. If you give them some time, they will not be as

harsh but will still retain a strong character. In fact, they will be delightful. If you find them still too hard when raw, use them in cooked desserts.

When you buy your currants in little baskets from the market, between the end of June and the beginning of August, the berries should be smooth and shiny, and, in the case of red currants, transparent. I prefer black currants with small, very black berries because they have more taste. Avoid crushed and overly ripe berries. If some of the berries are damaged, they will spoil the others in a short time. Very green and crisp leaves are a good sign. When tasting the perfect currant, the abundant juice will gush in your mouth, full of flavor, and will be refreshing without being too harsh.

Currants will last two to three days in the refrigerator in a sealed container. They should be left unwashed and should not be crammed together. Wash and dry them before separating the currants from the bunch. This will ensure that their juice does not wash away in the water. Do not freeze currants while still in a bunch. Defrost them slowly at room temperature, and their impeccable appearance will be restored.

Melons

From afar, a melon resembles a squash, and for a good reason: it belongs to the zucchini and pumpkin family. But what a difference on the inside! Its juicy and sweet flesh, a warm orange or a delicate green, will melt in your mouth. Melons are refreshing and energizing, and full of vitamins. There are numerous varieties, but there are five most commonly found at the market.

My favorite kind of melon is the one the French like the most, according to statistics: the popular Charentais variety known as the cantaloupe. This melon grows in France's largest agricultural regions—Vaucluse, Bouches-du-Rhône, Tarn-et-Garonne, and Lot-et-Garonne—where it is harvested from the beginning of June through the end of September, and is particularly abundant in July and August.

Two Charentais varieties are widespread: the one known as de Cavaillon or *charentais lisse* (smooth cantaloupe) is completely round and sleek. This melon is a light green that turns slightly yellow once ripe, and its well-marked grooves are an invitation to cut the fruit into wedges. Once the melon has attained the ideal amount of sugar, its orange-tinted flesh will have a honey undertone. Its relative fragility is its only fault. Sometimes more practical because it is sturdier, the *charentais brodé* (embroidered cantaloupe) is covered by a beige, corklike netting; its flesh and taste are very similar to that of the smooth variety.

Beneath the faint embroidery of the green-turning-yellow rind of the Galia melon hides a pale green and very sweet flesh. It is less refreshing than the cantaloupe. Originally from Israel, it has adapted well to France's Anjou region, where it is now cultivated.

The range of varieties have allowed the melon season to extend until November. The yellow Canary is an oblong melon from the south of France, with a lemon yellow rind, either smooth or textured. Inside, the very sweet flesh varies between a greenish white and orange. The melon named Olive Green, which is also oval in shape, has a green interior and a rind that is smooth or ribbed, yellow or green. Unlike other melons, its advantage is that it keeps for several weeks.

In order to properly choose a melon you have to smell it. It must exude a distinct fragrance, but one not too strong, which would be the sign that it is overly ripe. Melons do not improve if picked before they are ripe. No matter its variety, it should be surprisingly weighty. A good melon is heavy and dense, with a uniform, spotless rind. If the stem comes off or if there is some cracking at the base of the stem, the melon is good. But not all varieties will crack or have a stem, and the fruit might still be ripe. In France, some melons are sold with a small tag guaranteeing the perfect amount of sugar (between eleven and twelve percent). This information is very reliable. The abundance of juice and the flesh's silky texture should be a welcome surprise.

If well chosen, a melon can be kept for five or six days in a cool, well-ventilated area. It can be frozen peeled, sliced, topped with lemon or sugar, and wrapped in plastic. All melon varieties offer many possibilities, and can be enjoyed plain as an appetizer or dessert, or turned into jam or vinegar. Melons are also good in salads and prepared desserts. If you are going to eat the melon raw, serve it chilled, but not too cold or else you will miss most of its flavor.

Blueberries

I distinguish wild blueberries from cultivated ones. The former are small, fragile, and a bit tart, with a black inside that will stain everything they touch—but their flavor is marvelous. In France, you need only to hike in the Vosges, Jura, Cévennes, and Limousin regions in July to find them growing wild. Eat them immediately with a little sugar or cream. They also make sublime jams and tarts. Cultivated blueberries, prevalent in the United States, can be bought at the market between mid-July and mid-September. Plumper than the wild ones, they will resist the rigors of shipping and will keep for more than ten days at 50°F. Their blue skin with its white layer, the gauge of freshness, protects a white pulp that will not stain a thing. Despite their practical qualities, I do not like them as much as the wild variety. Because they are sweeter, they release less flavor. If removed from their branches and frozen as they are in a sealed container, the berries will revive well.

Blueberries are related to cranberries, and are also very similar to the small huckleberry, a slightly bitter forest fruit that is excellent in sauces for game dishes and your holiday turkey.

Plums

Among the four hundred varieties of plums, there are only three that interest me because I consider them the best. In August and September, I like to mix Reine Claudes, mirabelles, and damson plums in a bowl, creating a beautiful blend of yellow, green, and purple fruit tinted by a thin white layer. Made of fine specks of natural wax that keep the skin waterproof, this layer

is a sign of freshness and comes off easily when rubbed. Therefore, avoid plums that are too shiny, which means that they have been handled or stacked. These small, munchable treats do not last long. Take advantage of all the kinds at the market before it gets cold.

The Reine Claude is the queen of the table. These round plums, dressed in green with hints of yellow, pink, and sometimes purple, are available between July and the end of September. They have a golden flesh that is very flavorful, tender, and incredibly juicy. So sweet that they taste like honey, these plums make excellent preserves and tarts, which are best eaten warm. They are related to a variety called President, which comes along a bit later in the season. Very plump and oblong, a dark red or blue, Presidents are sometimes mistaken for quetsches but, with their yellow-green, juicy flesh, they are more like the Reine Claudes, if a little less sweet.

The mirabelle, the symbolic fruit of the Lorraine, also grows in other regions of France, such as Aveyron. A small ball of gold spotted with a crimson design, this fruit fills the senses with a distinct, extremely elegant fragrance, a sweet honey taste, and always a pleasant sour note. The pit releases a lot of flavor into this tiny fruit! Because they will cook without bursting, mirabelles work well in tarts and clafoutis. I also make very good preserves with them. Their flavor is embellished when sweetened with honey.

Damson plums, donned in a violet skin that verges on black, are firmer and have a more tart taste, which often means they are destined to be drowned in brandy or sunk into a simmering pot of jam. However, in Alsace, the region of France in which they are grown, their yellow-green flesh is delightfully placed in tarts and old-fashioned clafoutis.

As early as mid-June, the plump, multicolored newcomers, known as the Japanese-American varieties (Allo, Golden Japan, Black Amber, Santa Clara, Friar), make their appearance. They might impress you with their looks, but they still have not won me over. In terms of taste, they do not compare to France's traditional plums.

Choose your plums carefully. They should be nice and fat, without spots and wrinkles, and both firm and supple, neither too soft (meaning they are too ripe) nor too hard, indicating they were picked too early. Keep an eye on the forecast; their taste depends greatly on the whims of the weather and will seem more diluted after several days of rain. Once plums attain the optimum twelve percent sugar, they will offer a good blend of vitamins.

Buy plums as you need them. For lack of a cold room, they will keep for two or three days in your refrigerator's vegetable bin, but take them out at least one hour in advance so they are not too cold, which takes away their taste. If you wish to freeze them, remove their pits and place them immediately in a plastic bag in the freezer so that their pulp does not oxidize in the air.

Fall Fruit

Grapes

There are hundreds of varieties of grapes, and many are found at local markets in France. Among their incredible range of flavors the discriminating taster can detect green almond, caramel, acacia, rosemary, cherry, or hazelnut. But there are only four main types that you will find on a regular basis. All of them are very pulpy and filled with juice, their sweetness

always counterbalanced by a tinge of tartness. The muscat grape from Hamburg, which is small, black, and oblong, and has thin skin, often has the most flavor, with its typical musky taste. This variety can be found between the end of August and the beginning of November. Impatient grape lovers can find Chasselas grapes, which are available as early as mid-August and remain until the beginning of November; the small, round, and golden grapes have a thin skin, making them enjoyable to crunch. The Moissac Chasselas grown in France has a government-certified label of quality that guarantees the grapes were harvested at their perfect maturation. The Alphonse Lavellé grapes arrive soon after, with their big round fruit and thick black skin. The Italia variety is easily identifiable, with big, yellow, thin-skinned grapes that are both crunchy and juicy. As their name implies, they come from Italy, between September and December; in October, the same type is also harvested in the south of France and in the Pyrenees region.

If you do not want to peel them, choose grapes with thin skin for cooking. However, if you do plan to peel them, thick skin is easier to remove. Choose big grapes for quick cooking in a sauce. Thompson Seedless, imported from Chile, are very practical. In France, the National Institute for Agronomic Research (INRA) is working on similar varieties that offer substantial pulp and juice, yet save the cook from having to remove the seeds. Whatever the variety, choose heavy, consistently sized grapes that are both smooth and firm, and solidly attached to the vine. Fresh grapes should be covered with a thin white layer of natural wax, like plums. The stems and vines should be sturdy and full, showing no signs of drying.

Grapes fare especially poorly when dehydrated and last very little time in warm weather. To prolong them for a couple of days, remove any damaged grapes from the bunch as soon as you return from the market. They should be stored loosely in a container so that air can circulate or wrapped in perforated paper and placed very dry in the refrigerator vegetable bin. Remember to take them out an

Spices and Fruit

You will find that in my fruit desserts I use spices that are usually reserved for savory dishes.
It is common to add pepper, cinnamon, and cloves to the juices of cooking pears, peaches, or cherries.
But why not give other spices a whirl? It does not matter if the combinations seem unusual,
as long as they serve their purpose: either to bring out a certain quality or to soften an excess of fruit,
in order to arrive at the perfect balance of tastes. You are free to improvise,
but remember it is easier to add than to take away.

Gingerroot

Peeled and then thinly sliced into sticks, fresh ginger perfectly complements exotic fruit salads that include mango, pineapple, banana, and papaya. Ginger also goes well with all the melon varieties. Its spiciness freshens the palate.

Cinnamon

I prefer cinnamon sticks to powdered cinnamon. Cinnamon is best when blended with apples, pears, and plums, and its sweetness mixes exquisitely with that of vanilla.

Vanilla

These black, plump beans are our most familiar exotic spice. Choose supple vanilla beans and keep them packed together in a plastic bag in your freezer. Before using them, slice them in half, lengthwise, so that the fine seeds can release their sweet, sur-prising flavor. I also use vanilla extract, which gets the job done more quickly. Vanilla is extremely versatile: it can be used on virtually any type of fruit and in salads, compotes, tarts, desserts, syrups, and preserves.

Black Pepper

Black pepper plays an important role in cooking certain fruits—such as prunes, pears, and peaches—with wine. While cooking, tie the pepper-corns in a cheesecloth bundle, which you will remove once the flavor has been released. I also use ground pepper sprinkled over strawberries, pineapple, and melon. Make sure the pepper is ground finely.

Cloves

Like pepper, cloves can also be used when cooking fruit with wine. Let your imagination run wild. Grate cloves over a glass of lemon or orange juice, or stick the pungent buds into a lemon or orange rind and add it to freshly squeezed juice. It is up to you to make other discoveries.

Star Anise and Aniseed

The shores of the Mediterranean Sea are drenched in the familiar licorice aroma of aniseed, whether it comes from fennel, cumin, star anise, or green anise. Where I live in Provence, we scatter green anise in cookies, called navettes or croquants. I also use a few seeds in compotes, fruit salads, and syrups. Experiment with anise, keeping in mind that green anise, in small doses, is a particularly good complement to oranges, pineapples, figs, melons, papayas, and even strawberries. Fragrant and peppery star anise, named for its star-shaped buds, is often substituted for green anise. Its flavor is imposing, but re-freshing. Use star anise only on strong-tasting fruits (such as mango, melon, and rhubarb), whose taste can be counterbalanced by the spice.

hour before eating so that they return to room temperature. If grapes are too cold, they lose some of their flavor. Grapes separated from the bunch can be frozen in bags. During the winter you can toss them into fruit salads or sauces for game dishes.

Figs

I love figs for the same reasons I love the Mediterranean—they are seductively sweet, sumptuous, and show a wonderful contrast of colors. They also offer health benefits: figs are rich in calcium, magnesium, and vitamins.

There are many varieties of figs, which, for simplicity's sake, are grouped into two families: the more common black (or purple), and the green (or white). In theory, white figs are juicier than the purple ones. I personally do not think it is so simple; small or large, rounded or elongated like a pear, all figs can be good,

even though the varieties vary in sweetness. Their taste depends a lot on their maturity, and therefore on the number of hours in the sun, their region, and so forth. I do admit having a preference for the small black pear-shaped fig, which does not appear until the end of August.

There are also two kinds of figs distinguished by their season. Fig flowers ripen between mid-June and mid-July; they sprout the preceding summer, but do not develop until the following year. There are never a lot of them, and they are less sweet and more watery. Between the end of August and the beginning of November, the superb fall figs are harvested, having formed in the spring and ripened at the end of the summer. In the south of France, you do not have to look far to find a wild fig tree growing in the scrub.

Eat the thin-skinned varieties raw, reserving thick-skinned figs for cooking. The latter withstand the heat better and will not burst when baking or simmering briefly in sauces for duck, pork, rabbit, or guinea fowl.

Choose pulpy figs, which produce a drop of juice at the base and have small crevices on the surface of their skins. The fruit should barely resist a finger's pinch, but should not be too mushy. Its freshness can be judged by the stiffness of its stem. Figs that are picked too early will ripen a bit more, but will be less juicy.

Once ripe, figs will last only twenty-four hours after they are bought. If you want to keep them in the refrigerator for two or three days, store them in a sealed container, making sure they do not touch each other. Do not try freezing them; they will come out soft and flavorless.

Mulberries

Mulberries are the last of the smaller fruits to be harvested, not releasing sugar into their black-wrapped pulp until September or October. I prefer blackberries, which you can pick along nature's paths, as long as they have soaked up enough sun and have had enough time to ripen. Perfect berries overflow with a honey taste, which is a bit sharp because of the tannins, and their pigments will stain your mouth blue. When you cook them with sugar to make compotes, preserves, tarts, or prepared desserts, their silky, fruity flavor will greatly intensify. In France you rarely find mulberries and blackberries at the market. If they are small and shiny, and good and black, indulge!

You can freeze mulberries spread out on a plate, not touching. Once frozen hard, put them together in a plastic bag. Defrost them at room temperature, spreading them out once again.

Quinces

A fruit once forgotten by those unlucky enough not to grow them in their gardens, quinces are happily reappearing in markets. Do not try biting into this pear-shaped fruit, whose downy, shiny skin wavers between lemon yellow and gold when ripe; its hard, not-so-sweet flesh is so filled with sharp tannins that it will grate your tongue. Therefore, quinces are almost never eaten raw, except when cut into tiny pieces to wake up a salad—a salad with duck, for example.

But I love this unattractive fruit because of its extraordinary flavor. Quinces are generally found ripe at the market between the end of September and the beginning of November. If you grow them in your garden, make sure to gather them before the first frost, even if they are still green. Placed in a fruit bowl at room temperature, they will deliciously perfume your house for days, even weeks. A sturdy fruit, the quince will ripen slowly, lasting two or three weeks without spoiling as long as it is not bruised or a worm has not eaten into it. No use putting quinces in the refrigerator; the other food will be infected with their flavor.

Their strong, layered taste becomes more subtle with time. When ripe, the fuzz on the skin will simply rub off. Use quinces for compotes, preserves, aspics, or ratafia liqueur. My Aunt Célestine would give us quinces to stop diarrhea, a virtue officially recognized by nutrition specialists.

Winter Fruit

Apples

There are more than a thousand varieties of apples. What an orchard! Yellow, red, green, gray, or bicolor, apples are harvested and sent to market throughout the fall and winter seasons. The apple stand is never empty, although there is a limited choice in July and August. In France and America, each region can still brag about its traditional varieties, and many private gardens grow rare specimens. But only eighty-five varieties are produced regularly and only around fifteen are the pick of the basket. It would be impossible to describe each of them here in detail.

Picking out a fresh, ripe apple is easy. Choose a firm fruit with a tight skin. You will not be disappointed as long as there are not any blemishes.

To each his own apple. Make your selection according to personal taste, usage, and season. In France, our top stars appear roughly between October or November and April, except for the Rennet, which is around only in September and October; the Cox's Orange and Belle de Boskoop, which are gone as early as the end of February; and the Ida Red, which does not appear until January but lasts until June. The Gala is unique in that it is available between August and February, while the loyal Golden Delicious sticks around for all twelve months.

Golden Delicious apples do not deserve the bad reputation they sometimes endure. They can be among the best, but when subjugated to large-scale production, they can become bland. They are excellent when grown at a high altitude, on a hillside between 1,000 and 1,700 feet in elevation. Gray and white russets from Canada also grow in this way. They are identified in France by special names: Vigan Pippin and Savoy. In my recipes, I often suggest using Golden Delicious apples and pippins because they are readily available all year round and I like their taste. The former will not collapse when cooked, but the latter will. I therefore keep them for recipes that will not allow them to fall apart, such as tarts. When slices are placed on dough, they do not lose their shape.

Most apples are good to eat out of hand and offer a wide choice of flavors ranging from sweet to tart, with both crisp and soft consistencies, juicy and moist. The Belle de Boskoop, Golden Delicious, Melrose, Canadian russet, pippin, and Red Delicious varieties are best for cooking. For salads and other dishes in which the apples are to be used uncooked, opt for a Belle de Boskoop, Elstar, Granny Smith (especially for mousses), Ida Red, Canadian Russet, Gala, Golden Delicious (slightly unripe), or Jona Gold. I like blending the varieties in a compote; the firm pieces will remain crunchy while the others will melt into a purée, and the intermingled scents will result in a rich bouquet. Although they look hardy, apples do not withstand rough handling; they rot. So put them away carefully, preferably in a shaded, cool, and humid area where they will not dry out. If you want to keep them for one

or two weeks, put them in your vegetable bin in the refrigerator, but do not pile them up too much. However, they are eaten at room temperature. Although you might go through them quickly, there is no need to overcrowd your refrigerator with them. Apples can handle spending a few days at room temperature.

There is hardly reason to freeze apples since they are available year round. If you have a large harvest from your garden, peel and cut them into cubes or chunks for salads, and into thin slices for tarts. Squeeze some lemon over the pieces before putting them in a freezer bag.

Pears

There are even more varieties of pears than apples! France's National Institute for Agronomic Research (INRA) has counted as many as 1,500. Nine common types are enough to fill the markets all year long. This fruit is harvested throughout the seasons and comes from all over: from the Val-de-Loire and the south of France, as well as from Italy, South Africa, Chile, and Argentina.

Among the summer pears, the popular Guyot (or Doctor Jules Guyot, which is its real name) appears as early as July; ranging from a light green to a lemon yellow, it has a juicy, refreshing quality. The Williams is picked in August and eaten as early as September; everyone loves the juicy flesh of this big, stocky, sweet fruit, with its tart note, musky scent, and smooth, shiny skin. You can also find very beautiful ones splashed with red. Other similar varieties from

Dried Fruit

Dried fruits are emblematic of the tastes of the south of France. They are used in our reputed candy and pastries—in Calisons d'Aix, an iced marzipan sweet; in nougat; and in croquants, which are our famous hard cookies. Dried fruits can be found in the traditional thirteen desserts of Christmas, and compose the typical Provençal mendiant cookie, a symbolic blend (usually almonds, figs, hazelnuts, and Malaga raisins) that represents the robes of the four orders of Mendicant monks. In the past, dried fruits of all kinds were used abundantly in French savory dishes.

I cannot go into detail here about nuts and fruit in shells, like almonds, various walnuts, hazelnuts, pistachios, and pine nuts; or the many kinds of dried fruits, such as apricots, prunes, dates, raisins, figs, bananas, pears, and apples. However, there are numerous dried delicacies to suit everyone's taste.

Dried fruits should not be too dry; the quality of their taste often depends on their softness, or in the case of nuts, on their crunch. You should be able to enjoy them just as they are. Therefore, avoid products that are past their prime, which have lost too much water and are shriveled up. Also avoid nuts that have absorbed too much humidity and have, in turn, lost their crunch. Nuts can also go stale with time. Although the term dried implies that it can be kept for a long time, do not expect to keep the fruit many months before eating it. Choose dried fruits in sealed bags over those you buy in bulk, which are exposed to air.

Most regions that have a tradition of producing dried fruit want to maintain their reputation and are rigorous about quality. You can trust prunes from Agen, Tunisian or Algerian dates, Turkish figs and apricots, and Sicilian pistachios.

It is best not to choose raisins haphazardly; their flavor and scent will vary greatly depending on their label. The golden sultanas, which are tart and seedless, are similar to Smyrnas, which have a more musky taste. California raisins are very tender but not very flavorful. Corinth raisins are tiny and seedless, and have a typically fruity, subtle flavor. Malaga raisins are bigger and darker, but similar in taste to Smyrnas.

Keep dried fruit away from the sun, air, cold, and heat. Avoid refrigerating them, especially nuts, which soften with humidity. There is no need to freeze dried fruit since it is available year round.

When it comes to almonds, the word dried bothers me a bit because I especially like to use them after May, when almonds are picked fresh. At this point, when they are still green, tender, and soft, their shells shine a bit. The golden, woody shell can easily be split with a knife, and the light skin around the almond can be removed effortlessly. The nut is white, soft, tender, and milky. However, fresh almonds are hard to find outside of Provence. The dried kind will do the trick. We really only use the sweet almond; the bitter variety is toxic in large quantities and is used only in small doses in pastries.

---

the southern hemisphere are imported around February.

The Conference is the most typical of fall pears, distinguished by its long silhouette and its thick, rough skin tinted a soft brown with hints of red. Inside, the juicy pulp tastes of its long exposure to the summer sun. It is available between November and April. The popular Comice (or Anjou) pear is a large yellowish green fruit with thin, sleek skin that is softly painted with pink highlights and freckles. It has a subtle, silky flesh that is tart. Harvested in September, these pears are eaten in February or March.

The Passe-Crassane is the only winter pear grown in France; good and round, you will fine a buttery flesh beneath its thick, coarse, pale brown skin. Light and slightly sour, it can be found between November and April.

Other kinds of pears that appear in France are often more than a hundred years old, as their charming names will attest: the Louis-Bonne from Avrances, the Beurré Hardy, and the Alexandrine Douillard. Do not forget the small mountain pears that can be picked in the summer; although small, ugly, and so hard that they cannot be eaten raw, they are worth cooking. Their softened and concentrated flavor is a delicacy.

Pears cannot be eaten if they are not ripe enough. But we almost always buy them too hard, and that is normal. A pear is never edible upon picking. In particular, fall and winter pears have to go through a cold period before they can ripen. Our grandparents would pick them while

still green, then arrange them on racks in the cellar, waiting patiently for the skin to refine. Today, growers store them in cold, temperature-controlled rooms for a time before they are returned to room temperature to ripen. This process allows producers to space out maturation over many months. The pears that you buy, therefore, are in the midst of ripening. Keep them at room temperature, watching not to smash them together, and place them next to other ripened fruit if you want them to be ready sooner. The ethylene that is released will speed up the process. The later the variety, the longer the ripening time. A pear will become mushy very quickly after it reaches its optimum point. At the market, choose pears at different stages of maturity so that you can enjoy them day after day. Once a pear is ready to be eaten it can be kept in the refrigerator, but not for more than two days.

I do not peel the thin-skinned varieties, but do peel the gritty-skinned ones. I remove the hard bits that form near the core when the fruit has suffered from a difficult climate. The fall and winter pears will withstand cooking better. I cover them with lemon when I want to use them raw in a salad, which stops them from turning black. Before freezing, I poach them in boiling water with lemon.

Kiwis

Kiwis were practically unknown in France before World War II. At first they were an exotic fruit available only in fancy stores, but now they are quite common. France is the second largest European producer of kiwis, harvesting them from the end of November to mid-May in the south, where three-quarters of France's annual consumption takes place. This oval, downy fruit, brownish in color, isn't likely to seduce you with its exterior, but rather with the beauty of its inside—the bright green flesh, the white center and pale rays, and the little black seeds. With these qualities, kiwis have taken a prominent place in our desserts. Kiwis have a light flavor, but their slightly sour, refreshing flesh provides a tender

and pleasant texture. Beneath the skin, protected from oxygen, kiwis store a great supply of vitamin C, and their pulp is rich in citric acid.

Because there is really only one kiwi variety (Hayward) marketed in France, we are not overwhelmed with choices. Select kiwis with brown, tight skin and no bruises. It does not matter if they are not completely ripe. Kiwis will become tender if you keep them in an airy place at room temperature, as long as you do not pile other fruit on top of them. After a couple of days they will soften, while maintaining some firmness. At this point, they are perfectly ripe. Do not let them become mushy or they will no longer be edible. Once ripe, you can keep them for another forty-eight hours in your refrigerator's vegetable bin.

There is no need to freeze kiwis since they can be found year round. However, if your garden yields a great harvest, peel them, cut them into slices, and freeze them in small quantities in sealed bags.

Citrus Fruit

Oranges, grapefruits, lemons, limes, clementines, bergamots, citrons. These are some of the many members of the Rutaceae family in the genus *Citrus.* Although different in color, shape, and size, and ripening in various periods, they have a lot in common: citrus fruits grow along the same latitude and consist of a supple rind covering an inside separated into sections. With all the variations and origins, they are available at the market throughout the year and are all distinguished by their rich supply of vitamin C.

Pick citrus fruit with a smooth and firm rind, and without blemishes or dry, wrinkled areas. It should be compact and supple in your hand. A very green, small stem is the sign of freshness. Citrus fruits from the warmer climates might have a slightly green skin, which does not mean they are not ripe. They might even taste less tart than those with more color.

Citrus fruit can last a long time in your refrigerator's vegetable bin, but will not fare well in humid conditions, which will make them rot quickly. At room temperature or just slightly

chilled, you will enjoy their flavor more. In fact, it would be ideal to keep them in a cool room on an open tray so that air can circulate around the fruit. Turn them over now and then to make sure that they are not getting moldy.

If you wish to use the bitter zest of the rind for a dish or dessert, pick an organic fruit that has not been treated with chemicals and, if necessary, rinse it under hot water to dissolve the protective natural wax. No need to freeze citrus fruit since it is available all year round.

Oranges. Choose different varieties of oranges for eating, cooking, and squeezing. Do not consider the bitter bigarade for anything but cooking in salty dishes or for preparing orange wine. In any case, they are almost never found in markets, and bigarade lovers have to pick them in the south of France. In addition to local varieties, other oranges are imported from Spain, Morocco, South Africa, Tunisia, and Argentina.

Navel oranges are distinguished by their "umbilical" growth, which is pronounced in varying degrees at the top of the fruit. I prefer them for making desserts because they have practically no seeds. This is also the case with Washington navels, which are also very good for eating out of hand because of their crisp, not-too-juicy sections and sweet taste, as well as the rough-skinned early and late navel oranges.

Some types of navel oranges have neither a "button" up top nor seeds. With their generous amounts of juice, they are not always right for prepared desserts because they might drown them. However, I like to use them for juicing. From January to March you can find the oval Shamouti, the thin-skinned and round Salustiana, the oblong Valencia (or Morocco or Jaffa), which are more tart.

Finally, blood oranges, with their rinds tinted various shades of red, are less popular now than they were a couple of years ago. Perhaps this is because they are

more acidic than other types, and they have seeds, although they can be quite juicy. They appear between January and March.

Grapefruits and Pink Grapefruits. You can distinguish these two types by simply looking at them: grapefruits are slightly oval while a pink grapefruit is good and round. The grapefruit has a yellow rind, sometimes discretely marbled with green, and a sour-tasting pulp. The skin of a pink grapefruit is actually pink and protects a pretty interior that is dark yellow or pink, sometimes even red. Pink grapefruits are, in fact, the offspring of a real grapefruit and a Chinese orange. Pink or red flesh will always be sweeter than the yellow, although it will still have the fruit's characteristic bitterness.

Lemons and limes. Oh prolific lemon trees! Lemons grow only in regions where there is never a frost, and therefore no seasonal change. They flower and produce fruit several times a year. *Primofiore* lemons, or winter lemons, are harvested from October to December. The *limoni* appear from December to May. The *verdelli*, or summer lemons, arrive between June and September, and are often tinted with green due to extreme heat. Finally, the *bianquetto* or *mayolino* are the fall fruit.

Limes are not lemons but rather a citrus fruit from a slightly different botanical species *(Citrus latifolia)*. They are shaped like a small, seedless lemon and can provide even more juice. If you let them ripen on the tree, they will turn yellow!

There is no use getting mixed up by all the varieties; buy the fruit that is available when you need it. Their characteristics are very similar, although generally I find thin-skinned lemons to be a bit more juicy. Limes tend to dry up easily, so buy them in small quantities.

When whole, lemons and limes can keep for two weeks in the refrigerator. Halved lemons will last only two days wrapped in plastic.

Exotic Fruit

Bananas

A fruit without detectable seeds or a pit, and without a shell, bananas certainly accommodate our modern desires for ease and speed. In France they have become the third most widely consumed fruit, after apples and oranges. Meeting our needs for healthy snacking and quick meals, they are available all year round. Luckily, their soft, creamy texture and taste appeals to adults and children alike, and their flesh is rich in vitamins, magnesium, potassium, and iron.

Choose bananas without light brown spots, a sign that they have been frozen and will not ripen. Allow them to become completely ripe at room temperature, and definitely do not store them in the refrigerator, where they will turn black. On the other hand, they will ripen too quickly and will become mushy if stored over 68° F. To accelerate the ripening process, leave bananas in a well-ventillated area, wrapped in newspaper. Once good and ripe, bananas will have a strong yellow peel striped with black, and a lemon-vanilla aroma.

Almost all of the bananas sold in France come from Martinique, Cameroon, and the Ivory Coast, and belong to the Cavendish family. Poyo bananas, regular size and dwarf, have similar qualities, their size and curvature being the only difference.

I admit having a weakness for the pink banana, which grows mostly in Mexico and the Ivory Coast. Its ocher-pink or mahogany peel covers a soft fruit that is very sweet and flavorful; it should be supple in your hand. Finally, the dwarf banana or *dominico,* which grows in Mexico and the Antilles, has a solid, golden fruit beneath its thin yellow peel. It has a wonderful smell and is very sweet.

Very cold temperatures blacken and soften bananas and, therefore, we do not freeze them.

Pineapples

With their stylish leaves and unusual skin dotted with "eyes," the pineapple has an aristocratic air. It is the traditional symbol of hospitality. The "eyes" are, in fact, pulpy flowers that are joined together. Inside, the flesh is juicy and fibrous, with a refreshing sweet-sour taste that exudes a strong fragrance.

Among the five pineapple groups, there are really only two commonly found in markets between the end of October and mid-May. Those imported into France almost always come from the sunny Ivory Coast, and also from Cameroon, Brazil, Ghana, and the island of Réunion. The most common kind of pineapple is the smooth, quite cylindrical Cayenne, whose tinted orange exterior has no scales. Its pale yellow flesh has a very pretty scent.

The queen pineapple, or Queen Victoria, is smaller. Its golden rind, with its protruding eyes, surrounds a juicy, orange-tinted pulp with a tender core. Available in December and January, it is grown in Mauritius, Réunion, and Kenya. Although it is hard to find in France, I

really like Caribbean pineapple or red Spanish pineapple, with its light, peppery taste.

It is difficult to judge a pineapple's maturity from its color. Even if the eyes are evenly colored from the bottom to the top, a pineapple can be green on the outside and ripe on the inside! Rely, therefore, on the scent instead, which will develop in the final days of maturation. Choose the most fragrant fruit possible, but avoid any with even the smallest blemishes on the rind, a sign that it is overripe and has begun to ferment. It should be dense and heavy in your hand and should sport green, healthy leaves. At the peak of ripeness, pineapples are very rich in vitamins and enzymes that help in the digestion of proteins. Raw and chilled pineapples offer these benefits, but cooked or canned pineapples do not.

A pineapple will become completely ripe in a few days at room temperature. After that it can be kept in your refrigerator's vegetable bin for five or six days, but no more. Wrap it in paper so that it does not perfume the other produce.

To properly serve pineapple and to ensure the best flavor, cut it at the last second. Slice the fruit lengthwise to divide it evenly; the flesh is sweeter at the base.

To freeze pineapple, cut it into cubes. If you think you will be using them in sweet dishes, cover them with sugar.

Kumquats

Kumquats are part of the citrus family, but they are much less familiar than grapefruits and oranges. Tiny, about an inch in diameter, these fruit grow on trees that look like orange bonsai varieties. The kumquat can fit whole in your mouth and should never be peeled. The rind is deliciously sweet while the juicy pulp has a tart lemon flavor. I love feeling the fruit crunch between my teeth, letting the balance of flavors from the pulp and skin play on my palate. However, I do slice kumquats to remove their cumbersome seeds. The kumquat holds a place of honor in mixed fruit conserves.

Mangoes

Mangoes grow in the tropics, in all shapes and sizes: large or small, round or oval, plump or flat. Their thin, shiny skin, which is green before ripening, turns yellow, red, and even purple. The mangoes we find in our markets have a juicy, orange-colored flesh that is generally soft and buttery, and sweet but a bit tart. The mango has an especially delightful scent—it can smell like lemon, banana, peach, mint, and other surprising aromas. A large flat pit lies at its center.

Plump mangoes arrive at French markets almost all year round from Brazil, Mexico, and Peru (between November and March), or from western Africa (between March and July). Sometimes, in the winter, we enjoy mangoes from India, which are small and very tasty. I think they are exquisite.

In Asia and the Antilles, mangoes are used when still green, both raw and cooked, to accompany meat and fish. They are also mixed into curries and chutneys.

For desserts, it is better to use ripe mangoes. Pick one that is soft to the touch. Sliced in half, pitted, and simply scooped out with a small spoon, mangoes will provide iron and vitamins.

Mangoes should be frozen uncooked and peeled, sliced, or cubed. You will be able to use them in salads without further preparation.

Papayas

When their bright, green skin surrounds a still unripe fruit, papayas might almost be mistaken for large avocados, whether round, oval, or elongated. On closer inspection, you will notice that papayas are sleek and slightly ribbed. They become completely yellow when ripe. The skin covers a whitish flesh that turns a wonderful yellow-orange or red, depending on the variety. The flesh is juicy, flavorful, and refreshing, and the central cavity is filled with many black, bitter, and peppery seeds.

Whether they come from Brazil, the Ivory Coast, or Spain, papayas all have about the same taste. They are available all year long. I use unripe papayas in salty dishes, ripe ones for desserts. It is better to buy a papaya ripe—a bit soft to the touch but not too mushy. If it is still a bit hard, the papaya will ripen at room temperature. Finally, do not keep papayas in the refrigerator; rather, keep them chilled at about 55° F. They freeze well. Cut them beforehand into balls or chunks for adorning your salads.

Litchis

This thick, almost hard-skinned fruit, which is bright pink when ripe and punctuated with little bumps on the outside, will surprise you with its flavor: it tastes like roses mixed with muscat grapes. Once the shell is removed, you still have to remove the large stone, which takes up more room than the flesh. But litchis will make it up to you with their whitish, transparent pulp that is very sweet and seemingly without a sour note. It is, in fact, a fruit with one of the highest concentrations of sugar.

You will know a litchi is past its prime

if its shell and flesh have turned gray. Its flesh will also acquire a glassy quality.

The litchis we find in French markets are harvested mostly in Madagascar, and also in South Africa, the island of Réunion, and Mauritius, between November and March. Once picked they do not continue to ripen. Litchis will keep for four or five days, preferably in a cool, shaded area; light will turn their skin brown, making it brittle and susceptible to mold. Buy them in small quantities and eat them as soon as you can. However, they do often undergo a sulfur treatment (in proper amounts and not harmful) that prolongs their longevity for two or three weeks.

Do not freeze litchis. They will be too ugly when defrosted!

Tomatillos

This small, yellow-green fruit is wrapped in an oversized straw-colored calyx that looks likes paper. In France, tomatillos have pretty names like *amour en cage* (love in a cage) and winter cherry. The tomatillo came from South America, as did its relative the tomato. They now grow along the warm coasts of the Atlantic Ocean and the Mediterranean Sea, and are cultivated in South Africa, the United States, and a bit in France. If you live in the south, you can plant them in your garden. They grow easily and do not require a lot of care. Their tart taste is reminiscent of tomatoes and gooseberries, but I find the fruit to lack flavor. You can eat the whole thing: the pulp, of course, the thin skin and the small seeds. All you have to do is open the calyx, which separates into petals, and remove the fruit with your fingers. It will come out easily. I use tomatillos in their calyxes to decorate desserts, leaving them barely opened.

FRUIT SALADS AND COMPOTES

Paradise on earth is rustling in the depths of your garden. As a little boy, I'd find my happiness at the foot of a tree. I would first grab the fruits closest to the ground, and then I would shake the branches to make the ones I could not reach fall. Still not satisfied, I would finally climb the trunk up toward the most hidden temptations. Once, I remember—it was a very sunny day—I stuffed myself full of cherries; they were so warm they seemed much heavier in my stomach. I did not have the energy to climb down the tree. My parents found me later, half asleep, lying up there on a branch, blissfully happy. I had reached a state of fulfillment, in the true sense of the word.

I relive the pure sensuality of a plain piece of fruit, without crust or cream, every time I create a fruit salad. I bring together the most beautiful, ripe fruits, like raw materials, to make paintings and sculptures of colors and tastes. Sometimes I purée these plain fruits together or I soak them in a flavorful liquid: then the salad is called a soup.

If the fruits are not the prettiest specimens, if their cheeks are less chubby and not as smooth as they could be, and even if little black holes announce a worm within, I will not throw them out. "Insects have good taste and only choose sweet shelters," my Aunt Célestine would say. I make up for their imperfect ripeness; I hide their disappointing appearance by cooking them. Whole, cut into pieces, or blended together, they become compotes.

Six-Fruit Salad in Jellied Juice.
Recipe page 41.

FRUIT SALADS

A successful salad has everything to seduce you: the lush flavors of cut fresh fruit, the explosive colors of the exposed flesh, and the contrasting soft and crunchy textures that tickle the palate.

For perfect results, I select ripened fruit that is a bit firm, however, so that the pieces will not lose their shape. I leave small fruit whole (strawberries, cherries, litchis) and slice the others, respecting their natural form: rounded slices of banana and kiwi, thin peach wedges, halved apricots, and sections of citrus fruit. I make an exception for apples and pears, whose flesh will fall apart if sliced too thin: I therefore dice them or make balls with a melon-ball spoon.

I almost always squeeze lemon juice on sliced fruit. The acidity stops the oxidation process that turns them brown, while it also intensifies their fragrance and wakes up their flavor.

All fruits go well together in a salad as long as they are properly balanced. I almost always use even proportions.

Finally, I bring this patchwork to life with a personal touch: a few drops of alcohol (kirsch, Maraschino, Grand Marnier, Cointreau, champagne, fortified or syrupy wine, even red wine, if sweetened); or flowers, scented herbs, or a dash of spice (see pages 10, 16, and 22). Often, I top the salads with a scoop of an unusual flavor of homemade ice cream.

Once the salad is composed, I let it rest for a period of time before serving. I will often keep the salad in the refrigerator for thirty minutes to twenty-four hours, depending on the recipe, stirring it from time to time, but carefully so as not to damage the fruit. This allows the flavors to meld together in delicate harmony.

Peeling Citrus Fruit

To properly peel oranges, grapefruits, and lemons, you need to remove the rind and the inner white skin (known as the pith), as well as the thin membrane protecting the individual sections.

Place the fruit on a plate or cutting board with a groove to catch the juice. With a sharp knife, slice off a bit of the rind at the top and bottom, cutting to the pulp. Stand the fruit upright and cut away strips of peel from top to bottom, grazing the pulp each time.

To separate the sections, work over a colander set on top of a bowl to catch any juices. Using a knife, gently remove the seeds and discard. Slide the knife's blade along the inner edge of each section, moving outward from the center, and press the knife against the thin membrane to separate it from the pulp. Remove the sections.

Six-Fruit Salad in Jellied Juice
(Salade de Six Fruits en Gelée de Jus)

Photograph, page 38.

4 servings

Preparation: 30 minutes
Resting: 1 hour
Cooking: 5 minutes

1½ tablespoons gelatin
4 oranges
1 cup sweet white wine
  (such as Sauternes)
1 bouquet of fresh mint
1 apple
1 kiwi
1 peach
½ pound strawberries
½ pound raspberries

Create any kind of decoration beneath the jellied juice on the plate, using strawberry leaves, mint, or diced and candied zest. At the last moment, place flowers—such as nasturtiums, pansies, or jasmine—on the salad. They will enhance the freshness of the fruit.

In a small bowl, place the gelatin in ¾ cup cold water and let soften.

Extract the juice from 3 oranges. You should have about 1 cup of juice. Set aside.

In a medium saucepan, bring the wine to a boil. Boil for 2 to 3 minutes. Flambé and remove from heat. Add the orange juice and then the gelatin, stirring until it dissolves completely.

Strain the juice through a sieve. Chop about 10 mint leaves and add to the juice.

Let the mixture cool. Carefully pour into 4 large plates or shallow bowls. Refrigerate for one hour.

Peel the remaining orange and separate into sections (see page 40). Peel the apple, kiwi, and peach. Slice the apple and peach into wedges and the kiwi into rounded slices.

Set aside 4 pretty, whole strawberries. Wash the remaining strawberries, remove the stems, and slice into quarters.

Lightly rinse the raspberries and pat dry.

On each plate, arrange the cut fruit on top of the jellied juice. Distribute the raspberries evenly and garnish each plate with a sprig of mint and a strawberry in the center.

My Salad from the Moulin
(Ma Salade du Moulin)

8–10 servings

Preparation: 25 minutes
Resting: 1 to 2 hours
Cooking: 5 minutes

2 tablespoons sugar
2 tablespoons lavender
  honey

(continued on page 42)

This recipe takes its name from my restaurant because it is one of my favorite salads, with its eleven fruits representing every season and all parts of the world.

In a saucepan, heat the sugar, honey, vanilla bean, lemon juice, and ¾ cup of water.

Boil for 1 minute and remove from heat. Carefully remove the solids with a spoon. Let cool.

1 vanilla bean, split
    lengthwise
Juice of 1 lemon
2 bananas
2 peaches
2 nectarines
1 apple
1 pineapple
½ pound strawberries
¼ pound wild strawberries
¼ pound raspberries
2 oranges
1 pink grapefruit
⅔ pound fresh green
    almonds, shelled and
    peeled
Verbena leaves

Pour the cooled syrup in a bowl. Place a colander over the bowl and peel and section the oranges and grapefruit over the colander (see page 40). Remove the seeds.

Peel and slice the bananas, peaches, nectarines, apple, and pineapple. Stem the strawberries. In an attractive serving bowl, layer the fruit, finishing with the pineapple, and sprinkle with a few drops of lemon juice to prevent darkening. Refrigerate for 1 to 2 hours.

Just before serving, sprinkle the wild strawberries, raspberries, and fresh almonds on top. Garnish with a few verbena leaves. Leave the vanilla bean in the salad—there is always someone who will eat it!

Melon and Papaya Soup with Jasmine
(Soupe de Melon et de Papayes au Jasmin)

6 servings

Preparation: 20 minutes
Resting: 2 hours
No cooking required

Zest of 1 lemon
2 limes
1 ripe cantaloupe
2 ripe papayas
¾ cup sweet white wine
    (such as Sauternes)
⅓ cup sugar
1 pinch salt
36 very white jasmine
    flowers
6 sprigs of mint
White pepper, finely ground

Do not worry if you cannot find jasmine flowers. Even though the recipe will lose some of its charm, you will still enjoy it. The flowers are only for decoration and added fragrance. You can substitute orange blossoms. A few drops of orange blossom water drizzled on the fruit before blending would also work.

With a peeler, remove the zest of 1 lemon and place it in a small bowl. Squeeze the limes into the bowl.

Peel and seed the melon and papayas. Cut into chunks and purée in a food processor with the wine, sugar, salt, and the mixture of zest and lime juice. Pour into a bowl. Add 12 jasmine flowers and mix well. Refrigerate for at least 2 hours.

At serving time, strain the liquid through a sieve, mashing with a wooden spoon. Serve the soup in pretty bowls. Garnish each serving with 4 jasmine flowers and 1 sprig of mint. To finish, grind a dash of white pepper over each bowl.

Fruit Soup with Mint

(Soupe de Fruits à la Menthe)

If you like this zesty soup and find yourself making it time and again, you might want to play with it a bit. For example, substitute the mint leaves with verbena leaves. I have incorporated homemade vanilla sugar into my recipe (see page 7).

With a peeler, remove the zest from the lemons and oranges and julienne. Set aside.

Over a colander set on top of a bowl, peel the oranges and separate into sections (see page 40), reserving the juice. Add the juice of ½ of a lemon to the reserved orange juice. Set aside the orange sections.

Peel the ginger and slice into thin sticks or grate.

Prepare the candied zest. In a small saucepan, add half the orange zest, half the lemon zest, and the ginger, and cover with cold water. Bring to a boil for 1 minute. Drain in a sieve and cool immediately by running it under cold water.

Return the mixture to the saucepan and add cold water to cover, a few drops of lemon juice, 2 to 3 tablespoons of the reserved orange juice, the sugar, and the salt. Cook over low heat until all the liquid has evaporated. Be careful that the sugar does not caramelize. Remove from heat.

Meanwhile, remove the stems from the strawberries. Pit the cherries.

In a medium saucepan, heat the wine with the vanilla sugar, cinnamon, and remaining lemon and orange zest. Boil for 10 minutes over low heat. Add the strawberries and cherries. Remove from heat and let cool.

Peel the peaches and nectarines and slice into thin wedges. Cut the pineapple into small pieces. Divide the fruit in syrup and the cut fruit evenly into 4 shallow bowls. Refrigerate for 24 hours.

At serving time, garnish each bowl with the candied zest and a few sprigs of mint.

---

4 servings

Preparation: 35 minutes
Resting: 24 hours
Cooking: 20 minutes

2 lemons
2 oranges
4 ounces fresh grated
*   ginger, or 1 teaspoon*
*   powdered ginger*
1 tablespoon sugar
1 pinch salt
½ pound strawberries
1 cup cherries
1¼ cups full-bodied red
*   wine*
6 tablespoons vanilla sugar
¼ stick cinnamon
2 peaches
2 nectarines
A few slices canned
*   pineapple*
A few mint sprigs

Fruit Salad with Fruit Coulis and Vanilla Ice Cream

(Salade de Fruits au Coulis de Fruits et Glace Vanille)

4 servings

Preparation: 35 minutes
No cooking required

½ pound strawberries,
*   stems removed*
½ pound raspberries
Juice of 1 lemon
2 tablespoons confectioners'
*   sugar*
1 pink grapefruit
2 oranges
1 banana
1 kiwi
2 peaches
1 mango
4 scoops vanilla ice cream
A few verbena leaves or
*   lavender sprigs*

Do not blend the coulis too long: it will lighten in color and disturb the harmony of the intense palette. Prepare this salad as close to serving time as possible.

In a blender or food processor, blend half the strawberries and raspberries. Add ½ the lemon juice and the confectioners' sugar and blend again. Strain the coulis through a sieve. Refrigerate.

Over a colander set on top of a bowl, peel the grapefruit and oranges, and separate into sections (see page 40). Set aside. Add any reserved juice to the red fruit coulis.

Slice the banana. Peel and slice the kiwi. Peel the peaches and slice into thin wedges. Halve and pit the mango; peel and slice thinly. Sprinkle the fruit with the remaining lemon juice and set aside.

Coat the bottom of 4 shallow bowls or plates with the coulis. Arrange the fruit nicely on top, leaving an empty space in the center of each plate.

At serving time, place a big scoop of vanilla ice cream in the center. Garnish with the verbena leaves or lavender.

Pineapple Salad with Peppermint

(Salade d'Ananas à la Menthe Poivrée)

4 servings

Preparation: 15 minutes
Cooking: 2 minutes

2 pounds canned sliced
*   pineapple*
or 1 fresh pineapple, cut
*   into slices*
2 tablespoons light brown
*   sugar*
4 cherries or 4 pretty
*   strawberries*
4 sprigs + a few leaves fresh
*   peppermint*

I love the way peppermint brings finesse to the strong flavor of pineapple. This recipe is a delicious example.

Preheat the broiler.

Set aside 4 whole slices of pineapple. Cut the rest into half-inch pieces. Combine the cut-up pineapple and a few peppermint leaves. Divide evenly between 4 plates. Refrigerate.

Dust the 4 pineapple slices with light brown sugar. Put them under the broiler for 1 to 2 minutes to caramelize.

Remove the plates from the refrigerator and add a caramelized slice of pineapple to the center of each plate. Garnish the middle of each slice with 1 strawberry or cherry, and a sprig of peppermint.

Photograph, opposite.

Melon and Peach Zuppetta with Provençal Honey

(Zuppetta de Melon au Miel de ma Provence)

4 servings

Preparation: 25 minutes
Cooking: 1 minute

2 ripe cantaloupes
3 ripe peaches
1 vanilla bean
Juice of 1 lemon
2 tablespoons lavender
  honey
A few sprigs of mint

In Italian, zuppetta *means "small soup." Here we are reminded of the kinship between southern French and Italian cuisine, both of which reap the benefits of wonderful orchards. In the spring months, do what I do: garnish your* zuppetta *with wild strawberries. If you cannot find lavender honey from Provence, you will enjoy rosemary honey or wildflower honey. If you do not have a melon baller, cut the melon into small squares. Do not discard the scraped-out vanilla bean: use it to make vanilla sugar (see page 7).*

Halve the melons and remove their seeds. Using a melon-ball spoon, scoop out most of the flesh into little balls. Set aside.

With a big spoon, scrape out the rest of the melon flesh and put it in a blender or food processor.

Peel and pit the peaches. Slice 2 of them into wedges. Cut the third into large chunks and add to the melon.

With a small knife, slice the vanilla bean lengthwise. Scrape out the small seeds and add them to the melon and peach mixture. Purée.

In a small saucepan, heat the honey over a low flame just enough to liquefy it, about 1 minute. Add it to the fruit purée. Stir in the lemon juice.

Divide the soup evenly among 4 shallow bowls. In each bowl, arrange the melon balls in the center and the peach wedges around the outside. Garnish with a few sprigs of mint. Serve immediately.

César's Banana Salad with Wine and Honey

(Salade de Bananes de César au Vin et au Miel)

2 servings

I do not think my friend, the sculptor César, will be angry if I give away a recipe that he invented entirely on his own. His dessert shows just how good creations can be when they are simple. I could not hold myself back! Sometimes, we like to prepare mangoes in the exact same way. In either case, we always use lavender or wildflower honey.

Start no more than 30 minutes before serving. Otherwise, the bananas will turn black before they get to the table.

   Peel the bananas and slice. Place them in a bowl with the wine, honey, and a dash of finely ground pepper. Mix together. Serve immediately in 2 small dishes.

Preparation: 5 minutes
No cooking required

4 ripe bananas
1 cup good red wine
2 tablespoons lavender or
   wildflower honey
Finely ground pepper

Photograph, opposite. The salad is served with *fontainebleau* (a mixture of cream cheese and whipped cream) topped with honey.

Wild Strawberries and Nectarines with Rosé Champagne

(Fraises des Bois et Brugnons au Champagne Rosé)

6 servings

Here, the taste of wild strawberries and nectarines is intensified when fruit that has marinated a long time in syrup is combined with fresh fruit. In the end, the flavor of the raw fruit dominates the sweetened, slightly candied taste of the marinated fruit.

In a saucepan, place the sugar, vanilla sugar, and ³/₄ cup of water. Bring to a boil and remove from the heat. Add ¹/₂ the wild strawberries.

   Dice 2 nectarines (do not peel). Add to the syrup along with the lemon juice. Gently mix. Refrigerate 3 hours.

   At serving time, slice the remaining nectarine into wedges.

   Divide the syrup evenly among 6 shallow bowls. Top with champagne. Add the nectarine wedges and the remaining wild strawberries. Garnish with verbena sprigs. Serve very chilled.

Preparation: 15 minutes
Resting: 3 hours
Cooking: 5 minutes

3 tablespoons sugar
1 packet (or 2 teaspoons
   homemade) vanilla
   sugar (see page 7)
¹/₄ pound wild strawberries,
   stems removed
3 nectarines
Juice of 1 lemon
1 bottle of rosé champagne,
   chilled
Verbena leaves

White Peach Salad with Fresh Almonds

(Salade de Pêches Blanches aux Amandes Fraîches)

4 servings

Preparation: 15 minutes
Cooking: 5 minutes

2 tablespoons sugar
2 tablespoons lavender
    honey
1 vanilla bean, split
    lengthwise
Juice of 1 lemon
6 white peaches
2¼ pounds fresh, peeled
    almonds
Orange blossom water

One day, my friend César came up with this idea for a salad almost by chance, composing a dessert with what he had on hand. Luck was on his side: the combination of silky peaches and creamy fresh almonds was a real success. I only added my own touch to complete the recipe. Often I stand the filled bowl on a big round plate covered in fig leaves.

In a saucepan, heat 1 cup of water with the sugar, honey, vanilla bean, and lemon juice. Boil for a few seconds and remove from the heat. Remove the vanilla bean. Let cool.

Peel the peaches and cut into wedges. Gently stir them into the syrup with the fresh almonds.

Flavor with a few drops of orange blossom water. Pour into four bowls and serve immediately.

The best time for almonds is right after they are picked, when they are tender and creamy.

Blood Orange and Currant Aspic with Sauternes Wine

(Aspic d'Oranges Sanguines et de Groseilles au Sauternes)

This recipe is proof that marrying a great wine with fruit is delicious. Make sure the liquid does not boil, or else you will lose much of the Sauternes's rich flavor. You can serve this salad in small individual bowls instead of one big one. In any event, I love eating this dish with vanilla ice cream or chocolate mousse: what a symphony for the taste buds!

Prepare the candied orange zest. Using a peeler, remove the zest from 3 oranges. Julienne into thin strips. Stir the zest into a small saucepan of boiling water. Drain as soon as the water returns to the boil. Let cool.

Return the zest to the saucepan and add the sugar and 2 tablespoons of water. Heat over a low flame, stirring with a fork until the zest is covered by a shiny layer of syrup. The pieces should not stick together. Spread them out on a plate, keeping them separated. Set aside in a dry place.

Squeeze the three oranges into a colander set on top of a large bowl. Over the colander, peel the 7 remaining oranges and separate into sections (see page 40).

In a small bowl, place the gelatin in ½ cup cold water and let soften.

In a saucepan over medium heat, combine the reserved juice and the wine. Just before it boils, stir the gelatin into the saucepan until dissolved completely. Remove from the heat. Let cool completely, but do not let the syrup congeal.

Pour a thin layer of syrup into the bottom of a glass serving bowl. Layer the candied zest on top. Add a layer of orange sections and a few currants. Repeat each layer until the bowl is full. Try to avoid air pockets between the fruit. Refrigerate for 24 hours.

Serve on chilled plates so that the slightly congealed syrup will not melt immediately.

6–8 servings

Preparation: 40 minutes
Setting: 24 hours
Cooking: 10 minutes

10 organic blood oranges
4 tablespoons sugar
1 heaping teaspoon gelatin
⅔ cup Sauternes (or other sweet white wine)
½ pound currants, removed from their stems

Equipment
A transparent glass bowl

COMPOTES

Compotes are so easy to make! All you need to do is cook fruit in a syrup with varying amounts of sugar. Using this simple formula as a base, you can experiment all kinds of ways. I keep some fruits whole for the pleasure of feeling their texture. I cook fruits in their skins for more flavor, then I purée in a food mill, which also serves to remove the cooked skin.

I enjoy combining different fruits in the same compote. I make up for winter's avarice by using a mix of dried fruit.

Depending on the natural juice content of each fruit, I finish by mixing the fruit in the rich cooking syrup, or I use the syrup in other ways: in a fruit coulis as a sweetener, or boiled to make a thicker topping.

This cooking syrup can be enriched with a multitude of flavors: vanilla, cinnamon, mint, orange, lemon zest, cloves, almonds, coconut, and more. When flowers are in season, I never leave out the subtle taste that they can bring (see page 10). To satisfy my passion for the flavors of southern France, I often use honey (lavender, thyme, or rosemary) instead of sugar.

Whether served warm or cold, compotes deserve to be called a dessert when entertaining because they love company: I top them with whipped cream, adorn them with ice cream and cookies, powder them with vanilla sugar, or enhance them with spices (see page 7).

I also like surrounding the compote with little bowls of dried fruit or nuts, especially whole grilled hazelnuts. I sometimes use sour cream, yogurt, crème fraîche, or heavy cream as accompaniments. When it comes to dairy products, I love *fontainebleau*, which is a cream cheese that is completely drained, then whipped smooth and topped with an equal amount of whipped cream.

So prepare my easy compotes and reinvent them through the years. By playing with the flavors, you will create many of your own combinations.

Cherry Soup from Gordes
(Soupe de Cerises à la Façon de Gordes)

4–6 servings

Preparation: 45 minutes
Cooking: 15 minutes

1 bottle of wine from
  Côtes de Ventoux
1 cinnamon stick
1 vanilla bean, split
  lengthwise
2 stalks dried fennel
Zest of 1 lemon
3/4 cup sherry
1 cup sugar
2 tablespoons wildflower
  honey
2 tablespoons tapioca
2 1/2 pounds cherries (not
  pitted)

The village of Gordes, perched in the middle of the Vaucluse region, not far from the Cistercian abbey of Sénanque, prides itself on having two serious harvesting seasons: one for grapes and one for cherries. During the months of May and June, seasonal workers come from all over for the picking. Everything is still done by hand, even the arranging of the cherries in small crates so they will not be too crowded. Trucks, especially from Cavaillon, deliver them quickly to all the markets in France, Germany, and throughout Europe. I was often lucky enough to taste this soup that Céleste, the cook at the Château de Joucas near Gordes, would make with the local cherries. My friends who lived there received so many guests that Céleste's recipes enjoyed quite a lot of prestige. To take as much pleasure in this delicacy as I do, serve it chilled in small dishes with a raspberry sorbet and/or a kirsch or Maraschino ice cream.

In a saucepan, combine the wine, cinnamon, vanilla bean, fennel, and lemon zest. Simmer over moderate heat and remove from the heat as soon as it comes to a boil. Let infuse for 20 minutes. Strain through a sieve and return the liquid to the saucepan.

Add the sherry, sugar, and honey. Bring to a boil, then pour in the tapioca. Let boil for 5 minutes over a low flame.

Add the cherries. Return to a boil, then remove from the heat and cover.

Let cool to room temperature before refrigerating.

Rhubarb Compote with Rum and Ginger

(Compote de Rhubarbe Rhum-Gingembre)

5 servings

Preparation: 15 minutes
Cooking: 30 minutes

Fresh ginger, approximately
  ²/₃ cup sliced
2 cups sugar
1 vanilla bean, split
  lengthwise
Zest of ¹/₂ lemon
Zest of ¹/₂ orange
2¹/₄ pounds fresh rhubarb
¹/₄ cup rum

It is a pleasure eating this unusual and delicious compote with cookies or using it as filling for a tart. Sometimes I like to embellish its taste even more with a fresh touch: 5 minutes before it is done cooking, I add a few whole strawberries.

Peel the ginger and cut into thin slices. You should have about ²/₃ cup.

   In a large saucepan over moderate heat, combine 3 cups water, the sugar, the vanilla bean, the lemon and orange zests, and the sliced ginger. Mix well and bring to a boil. Remove from heat.

   At the same time, in another saucepan, bring 8¹/₂ cups of water to a boil. Peel the rhubarb stalks and cut into inch-long pieces. Blanch for 2–3 minutes in the boiling water. Drain in a colander and cool under the faucet.

   Add the rhubarb to the sugar syrup. Cook for 25–30 minutes, uncovered, over low heat, stirring occasionally, until the pieces are soft. Stir in the rum. Serve warm or cold.

Photograph, opposite.

Quince Compote

(Compote de Coings)

4–5 servings

Preparation: 30 minutes
Cooking: 40 minutes

2¹/₂ pounds quince
Juice of 1 lemon
1¹/₂ cups sugar
¹/₂ cup wildflower honey

This recipe will both soften and intensify the typically strong flavor that gushes from a quince. The taste becomes even more complex when I add honey.

Peel and core the quince. Cut into large chunks. Place the chunks in a bowl and coat with the lemon juice.

   Stir the quince into a saucepan of boiling water. When the water comes to a boil again, drain and cool the fruit in a colander under the faucet.

   In the saucepan, heat the sugar in ²/₃ cup of water. When it has dissolved, add the honey and the quince. Allow the mixture to cook slowly until the fruit is soft, about 25 minutes. Mix or blend in a food mill with a fine mesh sieve.

Cherry Compote with Kirsch and Almonds

(Compote de Cerises Kirsch-Amandes)

4 servings

Preparation: 15 minutes
Cooking: 15 minutes

2¼ pounds cherries
1½ cups sugar
1 vanilla bean, split
*   lengthwise*
2 heaping tablespoons
*   peeled whole almonds*
½ liqueur glass of kirsch

This is a simpler tasting recipe than the cherry soup from Gordes (see page 55) but will bring out the two unmistakable flavors of the cherry: the pulp and the pit.

Rinse the cherries.

In a saucepan, dissolve the sugar in ⅓ cup water and let boil for about 10 minutes. Add the vanilla bean, almonds, and cherries. Cook for another 5 minutes.

Drain the cherries in a colander placed over a serving bowl. Let the syrup cool. Stir the kirsch into the cooled syrup. Add the cherries.

Apricot Compote with Four Flavors

(Compote d'Abricots Quatre Parfums)

4 servings

Preparation: 30 minutes
Cooking: 10 minutes

5 tablespoons sugar
A scant ½ cup honey
2¼ pounds ripe apricots
2 vanilla beans, split
*   lengthwise*
A few verbena and peach
*   tree leaves, if available*

In winter, you can make this recipe with canned apricots. Choose apricots in a light syrup so that the final taste will not be too sweet. To simplify, you can eliminate one of the four flavors. The compote will be just slightly different, but still very good.

In a saucepan, combine the sugar, honey, and 8½ cups of water. Stir over moderate heat until the sugar dissolves, then boil for 5 minutes.

Add the whole apricots. Also add the vanilla beans and the verbena and peach tree leaves. When the syrup starts to simmer again, cook for no more than 4–5 minutes over low heat. Remove from the heat, cover the pan, and let cool. Drain and set the syrup aside.

Place the solids in a bowl. Remove the apricot pits (while cooking, the pits release a slight almond flavor). Scrape out the vanilla bean seeds with a small knife and discard the shells. Remove the leaves. Add the cooking syrup until it has a slightly gelled consistency.

Dried Fruit Compote
(Compote de Fruits Secs)

5–6 servings

Preparation: 10 minutes
Soaking: 2 hours
Cooking: 25–30 minutes

This is truly a classic recipe. When I want to emphasize the summer flavors, which have been stored up in the fruit over the winter, I use $^1/_2$ cup of honey instead of 1 cup of sugar. This nutritious compote can be made year-round, but you will appreciate it most when it is cold outside. For a Provençal Christmas dessert, serve this compote with traditional mendiant *cookies.*

Place the prunes, figs, and raisins in a bowl. Pour the boiling tea over the fruit. Let steep for about 2 hours.

In a saucepan, combine the wine, cinnamon, nutmeg, lemon and orange zest, cloves, vanilla bean, and sugar. Bring to a boil over high heat.

Drain the dried fruit and add it to the wine. Simmer for 20 minutes over a low flame. Serve warm or cold.

1 pound prunes
12 dried figs
$^1/_4$ pound ($^1/_3$ cup) golden
   raisins
$4^1/_2$ cups hot bergamot or
   Earl Grey tea
$4^1/_2$ cups good red wine
1 cinnamon stick
$^1/_2$ teaspoon nutmeg
Zest of $^1/_2$ lemon
Zest of $^1/_2$ orange
3 cloves
1 vanilla bean, split
   lengthwise
1 cup sugar

Available all winter long, dried fruits allow us to create recipes that are as energizing as they are diverse.

Strawberry Compote with Orange Blossom
(Compote de Fraises à la Fleur d'Oranger)

4 servings

Preparation: 10 minutes
Resting: 24 hours
Cooking: 5 minutes

2¹/₄ pounds strawberries
1 cup sugar
*1 tablespoon orange
    blossom water*

Instead of orange blossom water, you can cook a couple of orange blossoms or the zest of 1 orange and/or 1 vanilla bean split lengthwise with the water and sugar.

Quickly rinse the strawberries in a colander under the faucet. Dry and remove the stems. Place in a bowl.

In a saucepan, heat the sugar and ³/₄ cup of water. Boil for 5 minutes. Remove from the heat and add the orange blossom water. Pour the liquid over the strawberries. Refrigerate for 24 hours.

Orange
blossoms
on the
verge of
opening.

Fig Compote

(Compote de Figues)

The fig's elegant flavor needs nothing else. Therefore, this recipe is a rarity in my repertoire: I add neither spices nor honey, nor any other flavor foreign to the fruit. At the beginning of September I choose delicious Bellone figs, which grow only near Nice.

Rinse the figs in a colander under the faucet. Trim the pointed end and the small stem, if they are hard.

In a large saucepan, heat the sugar and $2/3$ cup of water. Boil for 5 minutes. Stir the figs into the boiling syrup. Reduce to low heat and poach for 6–8 minutes. Carefully drain the figs, reserving the cooking syrup in a bowl. Return the syrup to the saucepan and allow it to thicken over a high flame. Place the fruit in a bowl. Drizzle the thickened syrup over the fruit.

4 servings

Preparation: 10 minutes
Cooking: 15–20 minutes

$2^1/4$ pounds purple figs
$1^1/2$ cups sugar

Peach Compote with Three Flavors

(Compote de Pêches aux Trois Parfums)

I sometimes use lavender honey in this compote instead of the sugar.

Boil a large saucepan of water. Add the peaches and stir for 1 minute. Drain in a colander and cool under the faucet. Peel and cut the flesh into pieces.

Using a small knife, scrape out the vanilla bean seeds. Set aside.

In the saucepan over moderate heat, combine the sugar with $1^1/4$ cups of water. When dissolved, add the vanilla beans and their seeds, and the lemon and orange zest. Let simmer for 5 minutes. Add the peaches. Cook for 5 minutes more.

Drain the fruit, reserving the cooking syrup in a bowl. Remove the vanilla bean pods.

Place the fruit in a food processor and blend. Add the cooking syrup and purée until it has a slightly gelled consistency.

4–5 servings

Preparation: 20 minutes
Cooking: 11 minutes

$2^1/4$ pounds peaches
  (do not peel)
1 cup sugar
2 vanilla beans, split
  lengthwise
Zest of $1/2$ orange
Zest of $1/2$ lemon

Prune Compote with Wine and Tea

(Compote de Pruneaux au Vin et au Thé)

I enliven the taste of prunes in hundreds of ways. I add cinnamon sticks to the tea they soak in, or I let them absorb wine, which I boil with lemon or orange zest, cloves, star anise, or vanilla. Sometimes I replace the wine with flavored tea: apple, red fruit, orange, or bergamot. Below is one of the many variations. It is easy and delicious.

Photograph, opposite.

Rinse the prunes and soak them for about 2 hours in warm tea. When they are swollen, drain and place them in a saucepan. Cover with the wine. Add the sugar, vanilla bean, and lemon juice. Cook for about 30 minutes over low heat, stirring regularly. Serve warm or cold.

Preparation: 5 minutes
Soaking: 2 hours
Cooking: 30 minutes

1 pound prunes
2 cups light tea
2 cups red or white wine
1/2 cup sugar
1 vanilla bean, split
  lengthwise
1 tablespoon lemon juice

Mirabelle Plum Compote with Violets

(Compote de Mirabelles à la Violette)

4 servings

The taste of mirabelle plums is similar to that of honey, and therefore they go naturally well together. In an unexpected way, violets bring freshness and complexity, complementing this wonderful marriage of flavors.

Wash, dry, and pit the plums from the stem end without separating the halves.

In a saucepan, heat the honey in 2/3 cup of water. When it comes to a boil, add the violets. Remove from the heat and poach for 10 minutes.

Carefully remove the violets with a slotted spoon. Place the pan back over medium heat. When it boils, add the plums. Poach for 10–12 minutes. Pour into a bowl and serve.

Preparation: 20 minutes
Cooking and poaching:
  30 minutes

2 1/4 pounds mirabelle plums
A scant 1/2 cup honey
1 small bouquet of fresh
  violets

Apple Compote
(Compote de Pommes)

4 servings

Preparation: 10 minutes
Cooking: 20 minutes

2¼ pounds apples (about
    6 apples)
4 tablespoons vanilla sugar
    (see page 7)
Juice of ½ lemon

For best results, choose Golden Delicious apples, pippins, or russets. If you peel them, you will not be bothered by the skin getting in your teeth. However, if you leave the skins on, the compote will have more taste. I do not use a blender for this compote, as I would rather have whole chunks of apple on my plate. You will find that the less you do to this dish, the more you will enjoy it! You can even make the compote in a closed container in the microwave. Sometimes I fill a precooked crust with the compote to make a quick tart. Do as you please: you can use any of the homemade flavored sugars (see page 7) instead of the vanilla sugar, or you can add ½ stick of cinnamon, 1 star anise, or 1 piece of ginger to the fruit before cooking.

Quarter the apples and remove the cores and the seeds. Place the apples in a saucepan with the sugar and lemon. Cover and cook for 10 minutes over medium heat. Remove the cover and cook for another 10 minutes. Keep an eye on it to make sure the juice evaporates but that the fruit does not stick to the bottom.

Pour the compote into a pretty bowl and cover with plastic wrap. Refrigerate. It will keep up to three days.

Photograph, opposite.

Apple-Quince Compote
(Compote Pomme-Coing)

2–4 servings

Preparation: 20 minutes
Cooking: about 40 minutes

1 quince
1 tablespoon lemon juice
4 apples
1 tablespoon vanilla sugar
    (see page 7)
2 tablespoons sugar

For the holidays, add a small piece of ginger to the saucepan and serve the compote hot with warm foie gras, game, or duck.

Peel, quarter, and core the quince. Immediately coat the pieces with the lemon juice so they do not turn brown. Slice into thin strips.

Place the quince strips in a saucepan with 1½ cups water and the vanilla sugar. Cook for 15 minutes over medium-low heat.

Meanwhile, quarter and core the apples. Add them to the saucepan along with the sugar and cook over low heat, stirring occasionally, until the juice has evaporated, about 25 minutes.

TARTS
AND
TARTLETS

Among my earliest memories is the smell of a tart baking. That unmistakable aroma, one that blends the fragrances of sugar, warm fruit, and buttery crust, is one of life's greatest joys. When I make tarts now, I slip in a subtle, surprise flavor, which was my mother's touch, a whimsical ingredient that she would add even to the most traditional recipes. To recreate the pleasures of a homemade tart, always use pure butter, otherwise, you won't experience the same rich, tantalizing scent. It was butter that made me discover what a treat really was and convinced me that my future happiness lay in becoming a chef.

My mother would always roll out too much dough for the day's tart—I hoped there would be enough for another tart for the next day. If there was not quite enough for two, I would get a treat anyway. We would make little round pastries with the excess, or we would roll out the dough thickly and cut out special small cookies. We would mark crosses on the cookies with a knife or the tines of a fork, brush on some beaten egg, flavor with a drop of coffee, and put them in the oven. All the children who gathered around to share these little treats loved them—and the anticipation. We could not wait to devour the real tart!

It is true that if you only make one cake in your lifetime, it is usually a tart, no matter the season. Spring and summer are the best seasons for making tarts, but fall fruit can also yield great results. Any fruit that we are smart enough to preserve, sterilize, or freeze will be like a welcome dose of sun and warmth on a winter's day.

Two varieties of pippins: Canadian russets and Clochards for making soft apple tarts, a favorite dessert of the French.

PASTRY DOUGH AND TARTS

t is not surprising that beginners are drawn to tarts, which are easy to make, falling somewhere between fruit salads and more complicated desserts. Behind its simple, fresh appearance, a tart nevertheless requires careful attention. Chefs do not fool around when it comes to preparing tarts—a tart has its own rules that cannot be argued. It is both very simple and very precise.

The Rules of a Successful Tart

The fruit comes first! The quality of the fruit is the most important consideration. The fruit should also be the first thing you notice, so be generous: overlap the pieces.

A tart crust can have different textures, but it must always be made with butter. It should be thin enough that you almost do not notice it; its role is to enhance the fruit. If the crust dominates, you no longer have a tart. The extreme example is the thin-crust tart whose recipe you will find in this chapter.

You need to know what kind of crust is best for each fruit, and for some fruit you will need to alter the crust depending on the

way the fruit is sliced. If you slice an apple into thin strips, for example, you will need to choose a puff pastry dough, rolled out thinly. If you cut an apple into large sections, you will want a flaky, more rustic crust. The recipes in this chapter will offer you many examples that will allow you to improvise in the future.

The shape of the tart will change depending on whether or not you prebake without the fruit. For example, you might bury apricots, plums, cherries, pears, and apples in the dough, putting the dough up the sides of the pan to hold the fruit. On the other hand, if you are using strawberries, raspberries, or blueberries, make your life simple: bake a completely flat circle of dough, known as a "flat crust," and arrange the fruit, raw, on top.

The crust must be cooked uniformly on the bottom and edges. A tart pan made of thin metal with a removable bottom and a nonstick baking sheet are fine equipment for home use. Place the sheet or pan directly on a rack in the oven so that the bottom bakes uniformly from underneath. I do not like porcelain pans or pans made of other thick materials, which slow down the baking of the bottom crust. Baking time usually varies between 20 and 30 minutes, depending on the thickness of the tart, and the type and amount of fruit.

To finish, give the tart some shine. You can do so by brushing on a special coating used by pastry chefs that is made of a jelly hardened with fruit pectin (the same element that allows jams and jellies to congeal). This will produce a bright result, but I prefer a lighter fruit glaze made from heated and strained jam. Even though it should not be applied too early (it might liquefy after more than an hour), it often has better taste.

It is up to you now—express your every whim! Improvise like I do with flavor combinations of your own invention. Grind a little pepper over strawberries or cherries, dust pears with unsweetened cocoa powder, add a touch of wildflower honey to an apricot tart, or scatter crushed pistachios on a strawberry tart. You can make heart-shaped flat crusts for Valentine's Day, or cut them into other shapes, such as moons, stars, triangles, pine trees, or Santas.

And then there are other tricks. For example, between the crust and the fruit, create a hidden layer of almond paste, or add an extra-thin layer of sponge cake (*génoise*) on the baked crust before arranging the fruit so that it will soak up the juices. This is straightforward and easy. In general, however, be careful not to detract too much from the simplicity of a classic fruit tart by adding too many embellishments. If you cannot stop yourself, go to the next chapter on prepared desserts!

Pastry Dough for Fruit Tarts

In the following pages you will find the most traditional pastry dough for fruit tarts. You should not be surprised to learn that I often slip in an extra flavor here and there. For example, I mix powdered ginger into the flour when making a peach, plum, apricot, or pear tart; for an apple, raspberry, or orange tart, I add cinnamon instead.

Prepare your crust in advance and refrigerate it for two hours, wrapped in a dish towel or plastic wrap. Remove it thirty minutes before rolling it out.

Storing Your Unbaked Pie Crusts

Do not hesitate to make a tart because of the time it takes to prepare the dough. This problem is easily solved: make enough for several tarts at once. It will not be any more work! Divide the dough into several balls and wrap them in plastic. The dough will keep like this for eight to ten days in the refrigerator and for up to ten months in the freezer. If frozen, remove it one to two hours before rolling it out. You can also shape the dough in advance and freeze it in the pan.

Rolling Out the Dough

Flour your work surface, which should be a clean flat area. Also flour the rolling pin. It is useful to have a few pairs of wooden rods of different widths ($1/10$, $1/8$, $1/5$ inch). Roll the dough around the rods until you are able to evenly flatten the dough.

Shaping the Tart

Choose a tart pan with a removable bottom, which will allow you to unmold the tart without breaking it. Unless you are using a nonstick pan, grease it evenly with butter so that the dough will not stick. Roll the flattened dough in reverse onto the pin and unroll it into the pan (if you are baking fruit with the dough) or onto a nonstick baking sheet (for a flat crust). When working with a pan, press the dough lightly into the pan, lifting the edges of the dough and working it gently along the sides with your fingers. Roll the pin over the edges of the pan to trim any excess dough. Prick the bottom with a fork before filling it and/or baking. If necessary, refrigerate the dough while waiting for the oven to heat. (Cold is indispensable in maintaining the dough's appearance during baking.)

Prebaking Dough

When prebaking flat crusts without any filling, prick the dough and cover it with a circular piece of parchment paper that is slightly larger in diameter than the crust. Add a layer of dried beans on top to weight the crust while baking. Bake the crust for about fifteen minutes (in a preheated 425°F oven) until it turns a hazelnut color. Wait until the crust has cooled completely before removing it from the pan. Handle with care: it is fragile. If you prebake the dough many hours in advance, keep it in a dry area.

Filling the Tart

Once the oven has preheated, remove the dough from the refrigerator. To arrange the fruit, start with the edges and move toward the center. Respect the baking times in my recipes and your crusts will keep their shape without becoming soggy.

Proper Serving

Depending on the original shape—circular, square, rectangular, etc.—have fun cutting your tarts up into squares, strips, or triangles for serving. A serrated knife will cut the fragile tart without damaging it.

Short Pastry

(Pâte Brisée)

Yield: Two 8- or 9-inch
tart shells

2 cups flour
10 tablespoons butter,
    chilled and cut into pieces
3 tablespoons sugar
1 pinch salt
3 egg yolks
1/2 cup cold water

Whether or not you use a food processor, work the ingredients just long enough that they come together to form a ball—no longer. Otherwise, the dough will become tough and rubbery, and you will have problems rolling it out. Furthermore, it will harden while baking.

In a medium bowl, combine the flour, butter, sugar, and salt with your fingers until the mixture is blended—the butter should be in small bits the size of oatmeal flakes. In another bowl, combine the egg yolks and water. Add to the flour mixture and knead briefly. Bring dough together into a ball and wrap in a damp towel or plastic. Refrigerate.

You can also combine the same ingredients in a food processor in the same order, mixing them at medium speed just until the dough forms a ball and separates from the sides of the bowl. Be careful not to overwork the dough.

Sweet Short Pastry

(Pâte Sucrée)

Yield: Four 8- or 9-inch
tart shells

2 1/2 cups flour
1 1/4 cups butter, chilled and
    cut into pieces
1 cup sifted confectioners'
    sugar
1 pinch salt
1 whole egg
2 egg yolks

To sift the confectioners' sugar, put it in a sifter or sieve placed over a small bowl and use a spoon to break up any lumps.

In a large bowl, combine the flour, butter, sugar, and salt with your fingers or a dough hook until the butter is broken into bits the size of oatmeal flakes. Add the whole egg and the yolks. Blend for 1 to 2 minutes until the dough forms a ball. Wrap in a dish towel or plastic. Refrigerate for 30 minutes to 1 hour.

Sugar Pastry
(Pâte Sablée)

Yield: Three 8-inch tart shells

2 cups flour
¹/₂ cup sugar, or less depending on taste
Grated zest of 1 lemon
1 pinch salt
1 cup butter, chilled and cut into pieces
5 egg yolks

This is another recipe in which you should knead the dough as little as possible so that it retains its texture and elasticity.

In a medium bowl, combine the flour, sugar, lemon zest, and salt. Add the butter and the egg yolks. Knead briefly to combine. Form the dough into a ball. Wrap in a dish towel or plastic. Refrigerate for 2 hours or until firm.

Puff Pastry

(Pâte Feuilletée)

This is the most complicated dough to prepare. You cannot use a food processor. Everything must be done by hand, with care and precision. But in the end, what lightness, what crispness—what a delight!

1. Bind the dough with water. In a medium bowl, combine the flour and salt. Form a hole in the middle and fill it with the cold water and 7 table-spoons of slightly softened butter, cut into pieces. Blend rapidly with your fingers. Form a ball of dough and shape it into a square. With a knife score the top with a grid pattern. Wrap in plastic wrap and refrigerate for at least 30 minutes.

2. Work the butter. On a lightly floured board, roll out the rest of the butter into a 3 × 6-inch rectangle. Add a little flour to the rolling pin as needed. Refrigerate. For everything to work well, the dough and butter must be the same temperature.

3. Combine the dough and butter. On a floured pastry board, roll out the dough until you have a rectangle that is about 8 × 12 inches. Now place the butter rectangle in the center. Fold the edges of the dough completely over the butter, starting with the long sides and then the short sides, pressing the edges of the dough together. Roll out this pack-age until you have a new rectangle measuring approximately 20 inches in length and 8 inches in width. Fold it into thirds, folding back and under the short edges. Turn the board a quarter turn. Roll the dough out again until you have a rectangle of the same size. Again fold into thirds. Wrap the dough in plastic wrap and refrigerate for 20 to 30 minutes.

4. Repeat the above steps twice: roll out the dough, fold into thirds, let rest. Refrigerate at least 45 minutes or overnight, if desired. Just before using the dough, repeat the steps one more time.

5. Finish. Cut the amount you need. Roll out to the desired dimension. Wrap the leftover dough; it will keep for 2 to 3 days in the refrigerator.

Yield: Three 8-inch tarts

2½ cups sifted flour,
*   plus extra for flouring*
*   work surface*
1 teaspoon salt
1 pound butter:
*   7 tablespoons for*
*   binding and the rest*
*   for the butter layer,*
*   chilled*
1 cup cold water

Lemon Raspberry Tart
(Tarte au Citron et aux Framboises)

4 servings

Preparation: 25 minutes
Resting: 30 minutes
Baking: 25 minutes

⅓ recipe sugar pastry
  dough (see page 74)
3 scant tablespoons butter,
  softened
2 eggs
5 tablespoons sugar
Zest and juice of 1 lemon
½ pound raspberries
2 heaping tablespoons
  confectioners' sugar

Equipment
1 nonstick baking sheet

You'll love this tart even more if you serve it with a raspberry sorbet.

Place a medium bowl in the refrigerator to chill.

Roll out the dough with a pin. Cut out a circle 12 inches in diameter. Place it on an ungreased baking sheet and prick with a fork. Refrigerate for at least 30 minutes.

Preheat the oven to 350°F. Bake the dough for 20 minutes. Slide onto a rack and let cool.

Meanwhile, in a small saucepan away from heat, whip the butter until creamy. Add the eggs, sugar, lemon zest, and lemon juice. Place over low heat. Whisk constantly just until it comes to a boil.

Remove the pan from the heat and pour the pastry cream into the chilled bowl. Continue to whisk occasionally until completely cooled.

Pour the cream into the tart shell. Cover with raspberries.

At serving time, sift confectioners' sugar evenly over the raspberries.

Apricot Tart
(Tarte aux Abricots)

Photograph, opposite.

There are a hundred ways to make an apricot tart, but the recipe below is still my favorite. You can create variations: for example, replace the puff pastry with a flaky crust, over which you spread almond cream or sprinkle crushed macaroons; or use honey instead of sugar. In the off season, apricots in light syrup can be substituted for fresh fruit. Eat the tart warm, with vanilla ice cream or Beaumes-de-Venise muscat wine. This perfect blend of flavors will put you in a state of pure happiness.

Roll out the dough. Line the pan and sides with the dough, forming a high enough wall to hold the fruit. Prick the bottom with a fork. Refrigerate for 30 minutes.

Halve and pit the apricots.

Remove the dough from the refrigerator and arrange the apricot halves on top, standing them up and squeezing them one against the other. Sprinkle the blanched almonds over the top. Dust with sugar and refrigerate for 30 minutes.

Preheat the oven to 475°F. Remove tart from the refrigerator and bake for about 30 minutes. Although the oven is set very high, the fruit juice that is produced will protect the dough and bind the fruit in a delicious golden caramel syrup.

4 servings

Preparation: 30 minutes
Baking: 30 minutes

¹⁄₃ recipe puff pastry dough, or enough for an 8-inch tart (see page 75)
1 pound apricots
¹⁄₂ cup blanched almonds
1 cup sugar

Equipment
1 tart pan, 8 inches in diameter

Fig Tart with Cinnamon Crust and Honey Cream
(Tarte aux Figues en Croûte de Cannelle et Crème de Miel)

Fresh fig tarts are a staple in Provence. People in northern France rarely think to make them! There is no good reason why anyone should overlook this wonderful treat.

Roll out the dough and lay it in a tart pan. Prick the bottom with a fork. Refrigerate.

Preheat the oven to 375°F.

In a bowl, combine the ground almonds, 2 scant tablespoons of sugar, and cornstarch. Add the eggs and mix well.

In a saucepan, stir to combine the milk and honey and bring to a boil. Immediately pour into the sweetened egg mixture, whisking constantly.

4 servings

Preparation: 30 minutes
Baking: approximately 30 minutes

¹⁄₃ recipe sweet short pastry dough or sugar pastry dough, enough for a 9-inch tart shell (see pages 73 and 74), prepared with 1 teaspoon powdered cinnamon added to the flour

(continued on page 80)

¹/₄ cup blanched almonds,
    ground

2 scant tablespoons +
    1 tablespoon sugar

1 tablespoon cornstarch

2 whole eggs

¹/₄ cup milk

2 tablespoons wildflower
    honey

3¹/₂ tablespoons butter,
    softened

15 plump purple figs

4 heaping tablespoons
    confectioners' sugar

Equipment

1 tart pan, 9 inches in
    diameter

Return to the saucepan and bring to a boil, again whisking constantly.

Pour the cream into a clean bowl and whisk occasionally. When slightly cooled, add the butter and whisk until combined. Let cool completely.

Trim the fig stems and quarter the figs. Lay the quarters on a plate, open side up. Sprinkle with a tablespoon of sugar. Refrigerate until ready to use.

When the oven is hot, remove the chilled dough and figs from the refrigerator. Spread the cream on top and arrange the fig sections, flesh side up. Do not let sit. Bake for 20–25 minutes. Dust with sifted confectioners' sugar. Return to the oven for 5–10 minutes. Cool before serving.

---

4 servings

Preparation: 35 minutes
Baking: 25 minutes

¹/₃ recipe sweet short pastry
    dough (see page 73),
    prepared with 1 teaspoon
    powdered ginger added
    to the flour

7 tablespoons butter, softened

¹/₂ cup confectioners' sugar

¹/₂ cup blanched almonds,
    ground

3 ounces pistachio paste

2 eggs

1¹/₂ pounds strawberries

³/₄ cup apricot topping (see
    page 137)

¹/₄ cup pistachios, shelled

Equipment

1 tart pan, 8 or 9 inches
    in diameter

Strawberry Tart with Pistachio Cream
(Tarte aux Fraises à la Crème de Pistache)

With its cheerful red-and-green palette and superb flavor, this is one of the best recipes for a strawberry tart. Lovers of sweet treats are sure to go wild. The pistachio paste is indispensable: look for it at gourmet shops or ask your baker to provide you with some.

Preheat the oven to 350°F.

Roll out the dough and press it in the tart pan. Flute the edges with the tines of a fork or blunt edge of a knife. Prick the bottom with a fork. Place in the freezer for 10 minutes.

In the meantime, prepare the cream: by hand, beat the butter, sugar, and ground almonds. Add the pistachio paste and the eggs. Whip to a smooth, creamy texture.

Spread the cream on the dough. Bake for 15 minutes. Lower the temperature to 300°F and bake for another 10 minutes. Remove from the oven. Cool.

Rinse the strawberries, pat dry, and remove the stems. Halve if they are large. Fill the tart with strawberries.

In a saucepan over low heat, warm the apricot topping until it liquefies. Brush it over the strawberries.

Crush the pistachios coarsely, preferably in a food processor. Sprinkle over the tart.

Caramelized Pear Tart

(Tarte aux Poires Caramélisées)

4 servings

Preparation: 20 minutes
Baking: 20–25 minutes

4 ripe pears
Juice of 1 lemon
2 tablespoons pear brandy
5 tablespoons butter
¹/₂ cup sugar
¹/₃ recipe puff pastry or
  short pastry dough
  (see pages 73 and 75)

Equipment
A cast-iron skillet with
  high sides, or any frying
  pan 9¹/₂–10 inches in
  diameter with a
  heatproof handle

This recipe will remind you of a traditional tarte tatin. *Have you ever tried to make one with pears? Personally, I almost prefer the pear variation. If you want to stick to tradition, replace the pears with pippin apples and the pear brandy (which is optional) with Calvados.*

Peel the pears. Slice into 6 sections and remove the core and the stems. Sprinkle with the lemon juice and brandy.

Preheat the oven to 475°F.

Melt the butter in the skillet. Add the sugar, stirring, and remove from the heat when the mixture is smooth. Arrange the pear slices in the pan, forming a circle with their tips toward the center.

Transfer the pan to the oven and bake for about 20 minutes, until the pears are completely coated with the caramelized sugar.

In the meantime, roll out the dough into a circle equal in diameter to the pan. Prick with a fork. Once the pears are caramelized, remove the pan from the oven and cover with the dough, taking care to seal or fold along the sides. Lower the temperature to 350°F. Bake for another 15–20 minutes, until the crust is a beautiful golden color.

Remove the pan from the oven. Gently heat for 1 minute on the stove. Shake gently to make sure that the fruit is not sticking to the bottom. Wearing cooking mitts, turn a plate over onto the crust; in one quick movement, flip the pan over. The caramelized fruit should now appear on top. Serve warm.

Pear Tart with Cocoa

(Tarte aux Poires au Cacao)

Bitter cocoa powder does not contain any sugar at all, which makes it a striking complement to the silky sweetness of the pears.

Roll out the dough on a floured surface. Press the dough into the pan and prick the bottom with a fork. Refrigerate for 20 minutes.

Preheat the oven to 400°F.

Combine the cocoa and the grated chocolate in a bowl. In another bowl, mix the egg, sugar, and cream until it binds together.

Peel the pears and cut them in half. Core, stem, and slice thinly.

Remove the dough from the refrigerator. Sprinkle the bottom with the cocoa mixture. Arrange the pears slices on top, tips facing the center.

Pour the cream mixture over the pears. Bake for 20 minutes. The pears should turn a beautiful golden color. Remove from the pan and serve warm.

6 servings

Preparation: 25 minutes
Baking: 20 minutes

¹⁄₃ recipe sweet short pastry dough (see page 73)
1 heaping tablespoon bitter cocoa powder
3 ounces bakers' chocolate, finely grated
1 egg
2 heaping tablespoons sugar
¹⁄₃ cup heavy cream
3 ripe pears

Equipment
1 tart pan, 9 or 10 inches in diameter

Pear Tart with Chocolate Crust

(Tarte aux Poires sur Bisquit de Chocolat)

I cannot resist giving you a second recipe that combines pears and chocolate: in this one, there is chocolate in both the crust and the cream.

Prepare the dough. In a food processor, blend the flour, cocoa, confectioners' sugar, and butter until the dough adheres. Do not overblend.

Add the beaten egg and a pinch of salt. Process again. When the dough is blended and detaches from the sides of the bowl, form into a ball and wrap in plastic. Refrigerate for at least 1 hour.

Preheat the oven to 400°F.

Roll out the dough and fit it into a greased pan (or a nonstick pan). Prebake the dough for 20 minutes (see page 72). Cool for at least 30 minutes. Lower the oven temperature to 225°F.

Drain the pear halves well and arrange them on the tart bottom.

4–6 servings

Preparation: 30 minutes
Resting: 1 hour
Baking: 35 minutes

For the chocolate dough
³⁄₄ cup sifted flour
A scant 2 tablespoons bitter cocoa powder
¹⁄₄ cup confectioners' sugar
5 tablespoons butter
1 egg, beaten

Prepare the chocolate cream. In a saucepan, bring the cream and milk to a boil, stirring to blend. Remove from heat. Add the butter and chocolate and whisk until perfectly smooth. Add the beaten egg. While still hot, pour onto the pears. Bake for 10 minutes.

Remove from the oven and dust lightly with confectioners' sugar. Serve warm.

6 canned halved pears in syrup

For the chocolate cream
2/3 cup heavy cream
1/4 cup milk
1 1/2 tablespoons butter, chilled and cut into pieces
7 ounces grated chocolate (approximately 70% cocoa)
1 egg, beaten

Equipment
1 tart pan, 8 inches in diameter

Seven-Fruit Tart
(Tarte aux Sept Fruits)

6 servings

You can add polish to this spectacular dessert by garnishing it with 2 or 3 sugared currants (see page 171) and a few mint leaves on top of the dome. Serve with whipped cream. I make this tart year-round by varying the fruits: apples, pears, melon, pitted cherries, grapes, mangoes, papayas, bananas, mulberries, peaches, nectarines, plums—the choices are endless. Sometimes I make individual portions, using small bowls or coffee cups measuring 3 to 4 inches in diameter to cut out the dough.

Refrigerate a medium bowl.

Place the gelatin in 1 cup of cold water and let soften. In a saucepan, heat the apple juice and add the gelatin, stirring until it dissolves completely. Refrigerate only long enough to chill the mixture—do not let the juice gel.

Remove the chilled bowl from the refrigerator and drizzle a thin layer of the jellied juice along the bottom and sides. Cover with a layer of fruit. Continue adding layers of jellied juice and layers of fruit, mixing the colors, until you reach the top of the bowl. The last layer should be jellied juice. Refrigerate and let set for at least 6 hours.

Preparation: 35 minutes
Setting: 6 hours
Baking: 25 minutes

1 teaspoon gelatin
8 strawberries, quartered
2/3 cup apple juice
1/4 pound raspberries
1/4 pound currants, stems removed
2 apricots, quartered
1 small can of pineapple, drained and sliced into pieces

(continued on page 86)

1 orange, peeled and
    sectioned (see page 40)
1/4 recipe sweet short pastry
    dough (see page 73),
    prepared with 1 teaspoon
    powdered ginger added
    to the flour
4 tablespoons raspberry jam

Equipment
1 nonstick baking sheet
1 medium bowl with
    rounded bottom, 8 inches
    in diameter

Preheat the oven to 350°F. Roll out the dough into a circular shape about 1/8 inch thick and 8 inches in diameter. Prick with a fork. Transfer to a baking sheet and refrigerate for 1 hour. Bake for 20–25 minutes. As soon as you take it out of the oven, brush with the raspberry jam. Let cool.

Then dip the bottom of the fruit bowl in very hot water for 5 seconds. Dry it quickly and immediately invert it onto the crust. Serve at once.

Apple Tarts with Extra-Thin Crust
(Tartes aux Pommes Extra-Fines)

6 servings

Preparation: 30 minutes
Baking: 20 minutes

1/3 recipe puff pastry dough
    (see page 75)
3 Golden Delicious apples
2 tablespoons butter, melted
1 tablespoon sugar
1 pinch cinnamon

Equipment
1 or 2 nonstick baking sheets

It could not be simpler! To perfect these tarts, which should be eaten warm, serve with cinnamon or vanilla ice cream, or with whipped cream.

Roll out the dough as thinly as possible. Cut out 6 circles, each about 6 inches in diameter. Transfer to baking sheets. Refrigerate for 30 minutes.

Preheat the oven to 400°F.

Peel, halve, and core the apples. Slice thinly. Remove the dough from the refrigerator and cover with overlapping slices of apple.

In a small saucepan, melt the butter. Brush onto the tarts. Mix the sugar and cinnamon and sprinkle on top. Bake for about 20 minutes.

Puff Pastry Tartlets with Seasonal Fruit

(Tartelettes Feuilletées aux Fruits de Saison)

Preparation: 35 minutes
Resting: 1 hour
Baking: 30 minutes

¹/₂ recipe puff pastry dough
    (see page 75)
1 egg, beaten with
    1 tablespoon water
Confectioners' sugar

For the cream
2 scant tablespoons sugar
¹/₄ cup blanched almonds,
    ground
1 tablespoon cornstarch
2 eggs
¹/₄ cup milk
2 heaping tablespoons
    wildflower honey
3¹/₂ tablespoons butter,
    softened

For the fruit filling
6 strawberries, halved
Approximately 20 raspberries
1 pear, thickly sliced and
    sprinkled with lemon
1 peach, sliced into wedges
2 nectarines, sliced into
    wedges
6 tablespoons apricot jam
6 sugared currants (see
    page 171)
6 sprigs mint

Equipment
1 nonstick baking sheet

This recipe offers a spring/summer variation. You can substitute cherries or grapes for the raspberries, and apricots for the nectarines. Beyond that, feel free to alter the fruit selection based on what is available seasonally.

Prepare the cream. In a medium bowl, mix the sugar, ground almonds, and cornstarch. Add the eggs and mix well.

In a saucepan, heat the milk and honey, stirring to combine. When it comes to a boil, pour into the sugar-and-egg mixture, whisking constantly. When you have a smooth cream, return it to the saucepan. Bring to a boil, whisking constantly.

Pour the cream into a clean bowl. Whisk gently from time to time. Once it has cooled slightly, add the butter by bits and whisk to combine. Let cool.

Roll out the dough to a thickness of ¹/₈–¹/₄ inch. Cut out 6 4-inch squares. Cut the rest into strips measuring 3¹/₂ inches long and ¹/₂ inch wide.

Transfer the squares onto a baking sheet. Brush the edges with the egg beaten with 1 tablespoon of water. Lay dough strips along the edges of each square and pinch to join them at the ends. Stick a toothpick vertically into each side to hold the shape.

With the point of a small knife, score the edges of each tartlet with a chevron pattern. Alternately, you can press the dough into miniature tartlet pans with removable bottoms. Prick the center of each tartlet 2 or 3 times with a fork. Refrigerate for 1 hour.

Preheat the oven to 400°F.

Before baking the tartlets, brush the edges with the rest of the beaten egg. Bake for 15 minutes. Lower the temperature to 350°F and bake for another 10 minutes. Dust the edges with a thin layer of confectioners' sugar. Bake for 5 more minutes, until the edges caramelize. Do not allow to burn.

Fill each tartlet with 1 tablespoon of the cream. Decoratively arrange the fruit in each shell.

In a small saucepan, warm the jam over very low heat so that it liquefies. Strain through a sieve and spread onto the fruits with a brush. Let cool so that the glaze congeals. Garnish with the sugared currants and mint sprigs.

Pippin Apple Crumb Tart
(Croûte de Reinettes)

Photograph, opposite.

Like many tarts, this one is best when served warm, with vanilla ice cream.

Peel, quarter, and core the apples. Cut into ⅛-inch slices. Place in a bowl with the lemon juice, lemon zest, cinnamon, rum, and brown sugar. Carefully mix well and set aside.

Preheat the oven to 400°F.

Grease the pan with butter. Dust the bottom and the edges with the white sugar, tapping to remove any excess.

Spin the bread and butter in a food processor until you have fine crumbs. Divide into two batches.

Place half the bread-and-butter mixture in the pan, spreading it so that it extends up the sides. Pour in the apple mixture and distribute evenly. Cover with the remaining bread crumbs.

Bake for 15 minutes. Raise the temperature to 475°F and bake for another 15 minutes, until the crust is golden. Watch carefully.

With a small knife, detach the edges from the pan and invert the tart onto a plate, or serve directly from the casserole.

4 servings

Preparation: 30 minutes
Baking: 30 minutes

5 pippins or russets
Juice of 2 lemons
Zest of 1 lemon, finely grated
1 pinch cinnamon
2 tablespoons dark rum
⅓ cup brown sugar
8½ tablespoons butter, softened, plus additional to grease the pan
¼ cup sugar
Approx. ½ pound white bread, slightly stale, without the crust

Equipment
1 high-sided pan, 8 inches in diameter, or deep ovenproof casserole

Black Currant Crumb Tart
(Croûte au Cassis)

Set aside two black currants and crystallize them (see page 171) for a garnish. The time-saving aspect of this recipe is that the fruit, liqueur, and jellied sauce are combined and spread all at once onto the crust. It will also work with other small fruit, such as red currants, mulberries, and raspberries.

Preheat the oven to 350°F.

Roll out the dough with a rolling pin. Cut out 4 circles measuring 8 inches in diameter. Transfer to a baking sheet and prick with a fork. Refrigerate for 15 minutes.

Remove dough from the refrigerator and bake for 20 minutes. Slide onto a rack and let cool.

In a small saucepan, warm the black currant jam over very low heat until it liquefies. Remove from the heat. Add the crème de cassis and the black currants, which should be patted dry. Mix gently. Cool so that the mixture thickens again. Spread onto the cooled tartlets.

4 servings

Preparation: 10 minutes
Resting: 10–15 minutes
Baking: 20 minutes

⅓ recipe sweet short pastry or sugar pastry dough (see pages 73 and 74)
7 tablespoons black currant jam
2 tablespoons crème de cassis liqueur
½ pound black currants, rinsed and drained

Equipment
1 nonstick baking sheet

PREPARED DESSERTS AND BAKED GOODS

Every year, when the June sun shines, cherry trees, raspberry bushes, strawberries, and currants spatter their red palette throughout the garden. As summer continues, the orchards spill over with peaches, pears, apples, plums, hazelnuts, and quinces. During these fruitful days, which last through summer's end, my mother and aunt were always busy making traditional tarts, preserves, and compotes, as well as creating all kinds of recipes to celebrate the harvest. Admiring their incredible inventiveness, I came to understand the art of prepared desserts and baked goods.

Compared to fruit salads and tarts, prepared desserts seem more sophisticated, even if making them is not necessarily more difficult. You might get the impression of their complexity because of the countless ingredients involved, as well as the large variety of fruits—both whole fruits and those transformed into compotes, preserves, syrups, and coulis. If the fruit is not so pretty, you can hide it in a biscuit, a crêpe, a brioche; or bury it in a cream (a custard, a pastry cream, or a mousseline); coat it with caramel; or dunk it in a sauce.

Of the thousands of desserts that delighted and comforted me in my childhood, I liked clafoutis best, with its custardlike base infused with fruit, whether apricot, raspberry, or apple. I am especially nostalgic for the most traditional kind, made with small wild cherries that are as sweet as honey. I love this dessert even more since I was not allowed to have it as a very young child. My parents worried that the pits would ruin my teeth or that I would make a mess spitting them out.

They were right about one thing: you are not supposed to pit these cherries—besides, it is virtually impossible! While cooking, the pit will release a light almond taste reminiscent of kirsch. The best prepared desserts are sometimes those for which the search for good taste demands that we keep it simple.

Sweet Apricot Clafoutis with
Almond Milk. Recipe page 96.

Sweet Apricot Clafoutis with Almond Milk

(Clafoutis d'Abricots Mielleux au Lait d'Amandes)

Photograph, page 94.

8 servings

Preparation: 20 minutes
Resting: 1 hour
Cooking: 45 minutes

3 cups whole milk
1 cup blanched almonds, ground
5 tablespoons honey
$1/2$ teaspoon salt
7 eggs
5 tablespoons sifted flour
$3/4$ cup + 2 tablespoons sugar
7 tablespoons + 2 tablespoons butter
3 or 4 pounds fresh apricots, halved (do not pit)
$1/4$ cup kirsch

Equipment
1 tart pan, 12 inches in diameter

Use muscat apricots if you can find them (see pages 12 and 13). This clafoutis requires good quality kirsch. If you do not have any kirsch, or if you want a change, choose a bitter-tasting nut liqueur, perhaps the kind from Poissy, or an Amaretto. Do not leave your clafoutis to cool once it is cooked; it must be eaten warm. Serve with a golden Beaumes-de-Venise muscat wine. Chill the wine to 45–50°F. The dessert's bouquet intermingled with that of the wine will form a sumptuous and fruity symphony.

In a saucepan, bring the milk to a boil. Add the ground almonds. Remove from heat. Cover and let infuse for 1 hour. Stir in the honey and the salt while the milk is still warm.

Preheat the oven to 400°F.

In a small bowl, beat the eggs with a fork. In a large bowl, combine the flour and sugar. Gradually add the beaten eggs to the dry mixture, and then add the milk mixture, stirring constantly.

In a saucepan, heat 7 tablespoons of butter until it turns a light golden color. Pour immediately into the batter and mix well.

Using a brush, grease the tart pan with the 2 tablespoons butter. Sprinkle 2 tablespoons sugar in the pan so that it sticks to the bottom and sides. Arrange the apricots in the pan and cover with the batter.

Bake for 10 minutes. Lower the temperature to 350°F and bake for 30 minutes more.

Once the clafoutis comes out of the oven, sprinkle it with the kirsch. Transfer it from the pan onto a pretty plate while still warm.

My Wild Black Cherry Clafoutis

(Mon Clafoutis aux Cerises Noires Sauvages)

This is the famous clafoutis that made my mouth water all through my childhood. While making it, you will breathe in the delicious smell of the brown butter (beurre noisette). *It is absolutely necessary to make this with homemade vanilla sugar (see page 7). Go all the way and serve it with chilled whipped cream.*

Grease the pan with a generous amount of butter. Liberally dust the bottom and sides with sugar, then tap the pan to remove any excess sugar. Preheat the oven to 350°F.

Prepare the brown butter. Heat the butter in a small saucepan. When it starts to turn a hazelnut *(noisette)* color, remove the saucepan immediately from the heat. Filter through a small, fine strainer into a bowl and set aside.

Put the remaining ingredients, except the cherries, in a blender or a food processor. Blend to a purée.

Wash the cherries but do not pit. Dry them carefully with a dish towel. Pour them into the pan.

Briefly whisk the brown butter to emulsify and then pour it over the cherries.

Bake for 20 minutes. Lower the heat to 300°F and bake for 15 minutes more.

The clafoutis is done when its edges separate slightly from the pan or when a toothpick or knife inserted in the center comes out clean. Remove it from the oven and let it rest for 15 minutes. Transfer onto a pretty plate. Serve warm.

Preparation: 20 minutes
Cooking: 40 minutes

6 tablespoons butter, plus extra to grease the pan
2 teaspoons sugar to dust the pan
1/2 cup homemade vanilla sugar (see page 7)
1/2 cup blanched almonds, ground
4 eggs
Zest of 1 lemon, grated
Zest of 1 orange, grated
1/4 cup kirsch
1 cup milk
1 cup heavy cream or crème fraîche
1/2 cup cornstarch
1 pound small black wild cherries

Equipment
1 high-sided round pan, 10 inches in diameter

Sautéed Apricots

(Abricots Rôtis)

4 servings

Preparation and cooking:
20 minutes

³/₄ cup sugar
¹/₃ cup water
2¹/₄ pounds apricots,
* slightly ripe*
1 tablespoon wildflower
* honey*
3¹/₂ tablespoons butter
2 tablespoons slivered
* almonds*

I love to eat these richly browned apricots with a scoop of honey ice cream. Sometimes, after baking, I heat ¹/₄ cup kirsch in a small saucepan. Right before it boils, I pour it over the apricots and flambé. Then, as I take my first bite, I close my eyes to better savor the intoxicating tastes.

In a sauté pan (or in a medium-size deep pan), heat the sugar and water over low heat, stirring constantly. When the sugar has dissolved, turn the heat up to medium. Let cook without stirring until the syrup begins to turn a golden color.

Add the whole apricots and coat completely in the syrup, still over medium heat. Stir constantly for five minutes. Lower the heat if the caramel starts to darken too much.

Once the apricots have softened, stir in the honey and butter. For 2 to 3 minutes, roll the fruit in the bubbling juice until it is well covered in golden caramel. Pour into a bowl.

Spread the almonds in an ungreased nonstick pan. Lightly brown over medium heat, shaking regularly so the nuts do not burn. Remove from the pan immediately.

Sprinkle the almonds on the apricots and serve immediately. If you decide to flambé the fruit, add the almonds after you do so.

6–8 servings

Preparation: 25 minutes
Resting: 10 minutes
Baking: 40 minutes

For the caramel
¹/₂ cup sugar
1 tablespoon honey
1 tablespoon butter

12 slices canned pineapple
  in syrup

For the cake
¹/₂ pound (2 sticks) butter, at
  room temperature
1 cup sugar
3 eggs
1 cup sifted flour
¹/₄ ounce baking powder
¹/₃–¹/₂ cup canned pineapple
  juice

For the sauce
1 cup milk
3 bananas, cut in chunks
1 tablespoon vanilla sugar
  (see page 7)
¹/₄ cup rum

12 candied and halved
  bigarreau cherries
  (optional)

Equipment
1 high-sided pan, 10 inches
  in diameter

Pineapple Upside-Down Cake
(Biscuit Renversé à l'Ananas)

When I visited Jamaica, I enjoyed the cooking of a greatly talented Jamaican pastry chef, Mrs. Smith. She gave me marvelous recipes, including this upside-down biscuit, as it was known. I've kept it as a delicious memory that I am happy to share with you. Do like I do and eat this dessert with a coconut milk shake (all you have to do is blend milk with coconut ice cream). You can also make this recipe with apricots or pitted prunes. Soak them for 30 minutes in cold water and then drain them on a towel before starting.

Prepare the caramel. In a saucepan, combine the sugar, honey, and butter with 1 tablespoon of water. Over medium heat, cook the mixture until it turns a beautiful golden color (but not brown), stirring with a wooden spoon. Pour the caramel into a high-sided pan and let cool.

Drain the pineapple slices, reserving the juice to be added to the cake. Pat dry. Arrange them on the bottom of the pan.

Preheat the oven to 350°F.

Prepare the cake. Whisk the butter until creamy. Add the sugar and continue whisking until the mixture is thick and creamy. Add the eggs one by one, whisking constantly. Finally, add the sifted flour and the baking powder. Whisk lightly and add the pineapple juice.

Pour the batter into the pan and bake for 40 to 50 minutes. Check for doneness by inserting the tip of a knife into the cake, checking to see that it comes out clean. If there are traces of cake on the knife, bake a few minutes longer.

Remove the cake from the oven and let rest for 10 minutes. Meanwhile, prepare the sauce. Blend the milk, bananas, sugar, and rum in a food processor until emulsified and thickened. Serve on the side.

Place a plate upside down over the pan and in one quick movement flip the pan and plate over to invert the cake. The caramelized fruit will appear on top. Place 1 bigarreau cherry in the middle of each pineapple slice. Serve warm.

Bigarreau Cherry Cake with Black Pepper Cream

(Biscuit de Gros Bigarreaux, Crème de Poivre Noir)

6 servings

Preparation: 20 minutes for the cake, 15–20 minutes for the cream (including cooking)
Baking: 25–30 minutes

For the cake

1 cup (2 sticks) + 1½ tablespoons butter, softened

1 cup sugar

1 packet vanilla sugar or 2 teaspoons homemade (see page 7)

3 eggs, at room temperature

3 cups flour

1 heaping teaspoon baking powder

¼ cup orange juice

2 pounds cherries in syrup, drained

For the black pepper cream

1 cup milk

4 ground allspice berries

4 peppercorns

1 clove

4 egg yolks

4 tablespoons sugar

1 tablespoon heavy cream or crème fraîche

Equipment

1 large high-sided pan

This cake is easy to make, and the curious cream is sure to seduce you with its unusual taste—it is wonderful how the pepper complements the cherries! Allspice is not a pepper but a sweet spice that resembles a combination of cloves, nutmeg, and cinnamon. To really savor this dessert, eat it warm.

Prepare the cake. In a large bowl, whip the 2 sticks butter and the 2 sugars with an electric beater. Add the eggs one by one, beating constantly, until the mixture is foamy.

Sift the flour with the baking powder. Add to the mixture, continuing to beat. Finally, pour in the orange juice and blend.

Preheat the oven to 350°F.

With a brush, coat the pan with the remaining 1½ tablespoons butter. Dust with sugar, shaking the pan to distribute the sugar evenly on the bottom and sides. Flip the pan over on a table to release any excess sugar.

Pat the cherries dry. Arrange them on the bottom of the pan and pour the batter over them. Bake immediately for 25–30 minutes.

Meanwhile, prepare the black pepper cream. In a saucepan, bring the milk to a boil with the allspice, pepper, and clove. Remove from the heat, cover, and let infuse for 5 minutes.

In a bowl, beat the egg yolks and the sugar until you have a pale, frothy cream. Pour the hot milk into the egg-and-sugar mixture, stirring vigorously. Return to the saucepan and cook, stirring constantly with a wooden spoon, until the foam subsides. The cream should thicken. Remove immediately from the heat.

Add the heavy cream and strain the mixture through a strainer or sieve.

When the edges of the cake separate from the pan, remove from the oven and let rest for 10 minutes. Turn over onto a pretty plate. To serve, cut the cake into slices and pour the warm cream over each slice.

Green Grape Flan with Riesling Brandy

(Flan de Raisins Blancs au Marc de Riesling)

4 servings

Preparation: 20 minutes
Baking: 30 minutes

1 1/2 tablespoons butter,
    softened
4 tablespoons + 1 heaping
    tablespoon sugar
1 pound green grapes
2/3 cup flour
2 heaping tablespoons fine
    wheat flour
4 tablespoons blanched
    almonds, ground
1 pinch salt
2 whole eggs + 3 egg yolks
1 cup heavy cream or
    crème fraîche
1/4 cup Riesling brandy

Equipment
1 porcelain tart pan, 10 to
    12 inches in diameter

For this recipe, I always choose white muscat grapes, preferably seedless, if I can find them. But other varieties can be used, as long as the grapes are sweet and flavorful. Riesling brandy enlivens the taste; if you cannot find it, use muscat brandy instead. In any case, use a grape-based brandy.

Preheat the oven to 400°F.

Grease the pan with butter. Sprinkle with 1 tablespoon of sugar, shaking and tapping the pan to distribute evenly on the bottom and sides.

Wash, drain, and dry the grapes. Arrange them on the bottom of the pan.

In a medium bowl, combine the flour, wheat flour, 4 tablespoons sugar, ground almonds, and salt.

In another bowl, beat the whole eggs and the yolks. Blend in the cream. Combine with the dry mixture until the batter is smooth. Pour the cream mixture over the grapes.

Bake for 15 minutes. Lower the temperature to 300°F and bake for 15 minutes more. As soon as the flan comes out of the oven, sprinkle with the brandy. Serve warm.

Peach Mousse

(Mousse de Pêche)

4–6 servings

Preparation: 40 minutes
Cooling: 1 hour
Setting: 6 hours
Cooking: 25 minutes

2¼ pounds white peaches
Juice of 1 lemon
1½ cups + 3 heaping
    tablespoons sugar
3 eggs
2 teaspoons gelatin
⅓ cup peach liqueur
¾ cup heavy cream, chilled
Strawberry-Raspberry
    Coulis (see page 129)

With this recipe it is easy to slip in your own whimsical touch. For example, before letting the peeled peaches cool in the syrup, add a dozen or so verbena leaves. You will add charm to your dessert by choosing a mold in an unusual form, such as a heart or a clover shape. However, do not change the accompaniments; I always serve this mousse with a berry coulis. One more chef's trick: to speed up the setting of the mousse, place the mold in the refrigerator in a container filled with crushed ice.

In a large nonreactive saucepan, bring 3 quarts of water to a boil. Stir in the peaches and let boil for 2 minutes. Remove the peaches with a skimming ladle, and plunge in a large bowl of cold water. The skins will easily slide off.

Return the peaches to the saucepan and add the lemon juice, 1 cup sugar, and 4½ cups water. Bring to a boil. Lower the heat and simmer for 5 minutes. Remove from the heat and let cool.

Remove the peaches and reserve the syrup. Pit the fruit and place the pulp in a food processor. Blend to a purée, adding ½ cup of the reserved syrup.

Pour the purée into a saucepan with ½ cup sugar and 3 whole eggs. Heat over a low flame, beating constantly until thick.

Soften the gelatin in ½ cup cold water. Then stir the gelatin into the peach purée until it dissolves. Cook for another 1–2 minutes over low heat while stirring.

Add the peach liqueur. Remove from the heat and let cool, whisking the mixture from time to time.

Prepare the whipped cream. Pour the cream into a deep medium bowl. The cream should be very cold. Place this bowl into a larger bowl of water and ice. Mix with an electric beater until it forms stiff peaks. Add ¼ cup sugar while continuing to beat.

When the peach mixture has cooled completely, fold in the whipped cream with a spatula, stirring gently from bottom to top. Fill one big mold or several individual ones. Set for at least 6 hours in the refrigerator. Serve with a berry coulis (see page 129).

Apricot Mousse

(Mousse d'Abricot)

Proceed as with the peach mousse recipe, using the same proportions. However, it is not necessary to peel the apricots. Replace the peach liqueur with a nut-based liqueur, such as Noyau de Poissy or Amaretto, and serve with a strawberry coulis. I really love balancing these two fruits together because they share a similar tart note.

Pear Mousse

(Mousse de Poire)

Photograph, opposite.

Proceed as with the peach mousse recipe, using the same proportions, but peel the pears while raw and cut them into quarters before removing the cores and seeds. Substitute the same amount of pear brandy for the peach liqueur. Surround the mousse with a currant coulis (see page 129) and garnish with sugared currant berries (see page 170).

Roasted Figs with Nutmeg and Vanilla

(Figues Rôties à la Muscade et à la Vanille)

4–5 servings

Preparation: 15 minutes
Cooking: 30 minutes

4¹/₂ tablespoons butter
2 vanilla beans
12 ripe purple figs
1 whole nutmeg
*5 heaping tablespoons
    sugar*

In order to best savor these figs, serve them with a bowl of cold heavy cream or vanilla ice cream. You can also prepare peaches in the same manner, as well as pears, apricots, apples, and bananas. You only need to cook the apples and bananas for 20 minutes. Flavor apples with cinnamon or star anise, and bananas with star anise, sweetening them with light brown sugar.

Preheat the oven to 425°F.

Grease the bottom and sides of a baking dish with a generous amount of butter.

With a small knife, split the vanilla beans lengthwise and cut into ¹/₂-inch sticks.

Remove the fig stems. Halve the figs. Place them in a dish, packing them tightly, cut sides facing up. Scatter the vanilla-bean pieces over the figs. Grate the nutmeg over the entire dish and dust with the sugar. Bake for 15 minutes; lower the temperature to 325°F and bake for another 15 minutes. The fruit should be in a syrupy caramel.

Cool before serving.

Crispy Fig and Wild Strawberry Sabayon
(Gratin de Figues et de Fraises des Bois en Sabayon)

4 servings

Preparation: 20 minutes
Cooking: 7–8 minutes

8 plump Bellone figs
2 tablespoons + 1 teaspoon
*   light brown sugar*
Almond liqueur
¹/₃ pound wild strawberries
3 egg yolks
1 tablespoon red fruit
*   brandy*
1 tablespoon heavy cream
A few crystallized violets
*   (see page 170)*

Equipment
4 ovenproof plates

In Mougins, we call the little black figs from Nice, which are so flavorful and fragrant, Bellone figs. We consider them the best kind. If you cannot find them, choose small, very ripe figs. Depending on the season, play with the flavors and colors. For example, replace the violets with a few leaves of verbena and mint. In June, mix the figs with chopped fresh almonds; in the fall, chopped hazelnuts.

Rinse and dry the figs. Cut away the upper ¼ of each fig. Reserve the tops. If desired, peel the figs and cut into thick slices.

Place 1 fig top at the center of each of 4 plates and arrange the slices around it. Lightly sprinkle each with ½ teaspoon light brown sugar and a few drops of almond liqueur. Scatter a few wild strawberries over the top (save a couple for decoration).

Preheat the broiler.

In the top of a double boiler, combine the egg yolks, 1 tablespoon sugar, and the brandy. Set over simmering water on low heat. Beat until the mixture doubles in volume.

Remove from the heat and blend in the heavy cream. Carefully combine and pour onto the plates. Sprinkle each with ¼ tablespoon light brown sugar and brown for 2–3 minutes in the broiler.

Garnish with the remaining strawberries and crystallized violets. Serve immediately.

Russet Charlotte with Corinth Grapes

(Charlotte de Reinettes aux Raisins de Corinthe)

8 servings

Preparation: 30 minutes
Baking: approximately
  45 minutes

1 pound large brioche loaf
7 tablespoons melted butter
  + 6 tablespoons butter
1 vanilla bean
1 lemon
2¹/₂ pounds russet apples
¹/₄ cup Corinth grapes
5 tablespoons sugar
¹/₂ teaspoon cinnamon
Confectioners' sugar

Equipment
1 charlotte mold, 8 inches
  in diameter

Without ladyfingers or complicated creams, here is a rustic charlotte designed to enhance the fruit before all else. Nevertheless, you should serve it with a vanilla custard, offered warm like the cake. Do not heat it too much, it will go bad!

Preheat the broiler.

Remove the crust from the brioche loaf. Cut into slices and cut each slice on the diagonal to make triangles. Arrange on a baking sheet and place under the broiler until the slices turn a golden brown. Watch carefully. Do not flip the bread over and do not let it dry out too much. Butter the toasted sides while still hot with 7 tablespoons melted butter, stacking them as you go. Set aside.

Split the vanilla bean lengthwise and dice. Set aside. Zest the lemon; squeeze the juice. Set both aside.

Peel, quarter, and core the apples. Slice thinly. Sprinkle the slices with the lemon juice so that they do not turn brown.

Heat 6 tablespoons butter in a sauté pan. When it starts to turn golden brown, add the apples, then the sugar, cinnamon, and vanilla. Sauté over medium-high heat until the apples are tender. Add the lemon zest and grapes. Cook for another 1–2 minutes, stirring gently. Set aside.

Preheat the oven to 425°F.

Fill the bottom of a charlotte mold with 6 triangles of bread, buttered side down. Arrange the other triangles vertically around the whole mold, layering them slightly and making sure the buttered side is against the side of the mold.

Fill the mold with the apple mixture. Bake for 15 minutes. Lower the temperature to 350°F and bake 15 minutes more. Test for doneness.

Remove from the oven and let cool. Invert onto a flat plate.

Dust with confectioners' sugar.

Pears in a Dish with Browned Caramel

(Poires au Plat en Caramel Praliné)

6 servings

Preparation: 25 minutes
Cooking: approximately
35 minutes

12 pear halves, canned in
syrup
12 teaspoons hazelnut
praline paste
1 egg + 4 egg yolks
1 heaping tablespoon sugar
1¹/₃ cup heavy cream

For the caramel
¹/₂ cup sugar
¹/₃ cup heavy cream
2 tablespoons hazelnut
praline paste

Equipment
6 ovenproof ramekins

It is so easy to make a successful, creamy caramel! If you have some left over, you can keep it for a few days in the refrigerator and use it over ice cream. Ask your baker to make the praline paste or find it at a gourmet grocer.

Preheat the oven to 350°F.

Drain the pears. Fill each half with 1 teaspoon of praline paste. Place 2 pear halves in each of 6 ovenproof ramekins, rounded side facing up.

In a bowl, beat the egg yolks and the whole egg with the sugar. Add 1¹/₃ cups cream and combine. Pour over the pears.

Place the ramekins in a broiler pan. Pour water into the pan so that it is two-thirds full. Bake for 30 minutes. Remove the ramekins from the pan and let cool.

While the pears are cooking, prepare the caramel. In a high-sided saucepan, combine the sugar and 2 tablespoons water over medium heat. Cook until the sugar starts to brown. Remove from the heat and immediately pour in the ¹/₃ cup cream (make sure you wrap your hand in a towel to protect against splatters). Add the praline paste. Stir and set aside.

Drizzle a thin layer of caramel over each warm ramekin and serve immediately.

Orange Terrine

(Terrine à l'Orange)

6 servings

Preparation: 30 minutes
Setting: 2 hours minimum
Cooking time: 1 minute

⅔ *cup sweet white wine*
(such as Sauternes)
⅔ *cup orange juice (the*
juice of approximately
2 squeezed oranges)
½ *stick cinnamon*
½ *vanilla bean*
3 cloves
¼ *cup sugar*
1 tablespoon gelatin
8 oranges

Equipment

1 rectangular terrine mold
or loaf pan

The warm golden color and refreshing taste of this lightly spiced dessert, right for all seasons, is sure to please your guests. If you have time, perfect the presentation: save some liquid gelatin and pour it around the dessert once it has been removed from its mold. Let it set in the refrigerator.

In a saucepan, heat the wine with the orange juice and boil for 5 minutes. Remove from the heat. Add the cinnamon, vanilla, and cloves. Cover and let infuse for 15 minutes.

Soften the gelatin in ¼–½ cup cold water.

With a slotted spoon or strainer, remove the spices. Add the sugar and gelatin to the warm liquid and stir until the gelatin dissolves. If necessary, reheat it a bit so that it dissolves completely. Let cool to room temperature.

Peel the oranges and separate into sections (see page 40).

Pour a layer of the jellied liquid into the bottom of the terrine. Add a layer of orange sections. Alternate layers of jellied liquid and orange sections until you have reached the top of the terrine.

Refrigerate to set, at least 2 hours. To serve, dip the bottom of the terrine in hot water for a couple of seconds and then invert it onto a serving plate.

Fruit Fritters
(Beignets de Fruits)

When I was a child, we were overjoyed to have "dessert dinners" composed around fritters of different fruit, rolled in sugar. The secret to good fritters lies in the batter. It should be light, with a subtle flavor that does not compete with the fruit. The crispy outside must only accent the fruit's sweet taste.

Preparing the Fruit

The principle is almost always the same: cut the fruit; lay it out on a plate or tray; dust it with sugar and sprinkle it with alcohol; and then cover it, preferably with plastic wrap, and marinate before dipping it into the batter. Add chopped mint leaves, verbena leaves, or fresh flowers—orange blossoms, jasmine, violets—to the marinade.

Frying the Fritters

When the fruit has finished marinating (in the following pages, see the recipes on how to prepare each fruit), skewer each piece onto a toothpick and dip it into the batter, covering it completely. Then lay it into the oil. Use a subtly flavored vegetable oil—a corn oil, for example. Never heat it to more than 350°F. Make sure the fritters do not touch, as they might stick together. Flip them over once while frying, when the bottom is golden. Strain your fritters when the second side is browned, but do not use a skimmer; too much oil will be retained. Instead, use a wire spoon, which is similar in shape but has larger openings. Drain the fritters briefly on paper towels. Dust with confectioners' sugar or granulated sugar and serve immediately.

Banana Fritters (recipe page 120), served with a mango coulis (recipe page 129).

Batter
(Pâte à Frire)

6–8 servings

Preparation: 10 minutes
Resting: 30 minutes

2 cups flour
2 whole eggs
3 egg whites
2 pinches salt
¼ cup oil
¾ cup lager beer, chilled

You can always add allure to your fritters by directly flavoring the batter. Mix in some ginger, either freshly grated or dried and powdered, or some finely grated lemon or orange zest. You can also stir in star anise or powdered cinnamon.

Put the flour in a medium bowl and make a well in the center. Fill with the whole eggs, a pinch of salt, oil, and beer. Whisk, starting in the center, and incorporating the flour a little at a time. When blended, let rest for 30 minutes, covered.

Before using the batter, beat the egg whites and a pinch of salt with a whisk or electric beater to form stiff peaks. Carefully fold into the batter, stirring from bottom to top.

Apple Fritters
(Beignets aux Pommes)

6–8 servings

1½ pounds Golden Delicious apples
¼ cup sugar
¼ cup Calvados or Cognac

These are the most classic kind of fritters. I love to eat them with apricot jam. For children, you can always replace the alcohol with orange blossom water.

Photograph, opposite.

Peel, core, and slice the apples into ¼-inch wedges (allow for 3 slices per person for a dessert).

Place the slices on a tray or plate. Dust them with sugar and sprinkle with Calvados (apple brandy) or Cognac. Turn to soak both sides. Cover and refrigerate for 2 hours.

Proceed with batter recipe (above) and frying instructions (page 116).

Prune Fritters Soaked in Armagnac
(Beignets de Pruneaux Gonflés à l'Armagnac)

6–8 servings

1 vanilla bean
1 cup sugar
⅔ cup Armagnac
1 pound pitted prunes

Split the vanilla bean in half, lengthwise.

In a saucepan, combine 1 cup water, the sugar, and the vanilla bean. Bring to a boil for 1 minute. Pour into a medium bowl. Add the Armagnac and then the prunes. Let marinate for 1 hour. Drain well.

Proceed with batter recipe (above) and frying instructions (page 116).

Apricot Fritters

(Beignets d'Abricots)

4 servings

6 apricots
3 tablespoons sugar
2 tablespoons kirsch

The best apricots are the orange-colored variety from Provence, which are firm and flavorful. Serve these fritters with a fresh apricot coulis.

Halve the apricots and pit them. Place them in a dish, cut side up. Dust with sugar and sprinkle with kirsch. Cover and let marinate for 2 hours.
    Proceed with batter recipe (page 118) and frying instructions (page 116).

Pineapple Fritters

(Beignets d'Ananas)

4 servings

1 fresh pineapple or a small
    can of sliced pineapple
3 tablespoons sugar
¹/₄ cup Grand Marnier, rum,
    or kirsch

I leave it up to you to choose the alcohol in which to soak the fruit: Grand Marnier, rum, or kirsch.

Remove the pineapple skin and slice. Core with a wide corer. Put the slices on a large plate or tray. Dust with sugar and douse with alcohol. Cover and let marinate for 2 hours.
    Proceed with batter recipe (page 118) and frying instructions (page 116).

Banana Fritters

(Beignets de Banane)

4 servings

2 bananas
2 tablespoons rum
1 tablespoon sugar

A fresh mango coulis will heighten the flavor of these tasty fritters. You'll find the recipe on page 129.

Photograph,
page 117.

Peel the bananas. Cut them in half, lengthwise, or into thick slices. Sprinkle with rum and dust with sugar. Cover and let marinate for 2 hours.
    Proceed with batter recipe (page 118) and frying instructions (page 116).

My Mother's Apple Rice Pudding

(Gâteau de Pommes au Riz de ma Mère)

My mother would make this recipe with small, very flavorful apples and Italian short-grain Arborio rice, the creamy type of rice used to make risotto. While the rice cooks, the outside of the grains softens while the inside remains a bit firm.

Preheat the oven to 425°F. Grease the bottom of a baking dish with the butter.

Peel the apples, cut in half, and core. Pack the apples in the baking dish. Sprinkle ½ cup sugar on top. Bake for about 25 minutes, until the apples are perfectly caramelized. Check often to avoid burning. Remove the dish and reduce oven temperature to 350°F.

In a saucepan, combine 8½ cups water and the salt. Bring to a boil and stir in the rice. Once it boils again, reduce heat to medium and cook for eight minutes, stirring occasionally. Drain the rice but do not cool.

In another saucepan, boil the milk with the lemon zest, vanilla bean (split lengthwise), and ½ cup sugar. Add the rice. Cook for another 5 minutes, stirring regularly. Pour the mixture onto the apples. Return the dish to the oven for about 15 minutes, checking for doneness. Serve warm.

4–6 servings

Preparation: 25 minutes
Cooking: 55 minutes

6 tablespoons butter
3½ pounds pippin apples
1 cup sugar
1 teaspoon salt
¾ cup Arborio rice
4½ cups milk
Zest of 1 lemon
1 vanilla bean

Equipment
A Pyrex or earthenware baking dish, 8–10 inches long

Fresh Currant Bread Pudding

(Pudding de Groseilles Fraîches)

For this delicious pudding I suggest using the inside of a single large brioche. You can use the rest of the bread for breakfast the next day. But do not count on having any bread pudding left over—that never happens!

Preheat the oven to 400°F. Grease the sides and bottom of a high-sided pan with a generous amount of butter. Sprinkle with 2 tablespoons sugar, tapping and shaking the pan to distribute evenly. Set aside.

Slice the brioche into ¾-inch cubes. Lay the pieces on a baking sheet and place in the oven for about 5–10 minutes until dry, but not brown.

In a bowl, mix the eggs, yolks, and ½ cup sugar, then add the cream.

4–6 servings

Preparation: 25 minutes
Resting: 30 minutes
Cooking: 35 minutes

1½ tablespoons butter, softened
½ cup + 2 tablespoons sugar

(continued on page 122)

12 ounces brioche, crust
    removed
1 pound fresh currants,
    stems removed
2 whole eggs
2 egg yolks
³/₄ cup heavy cream or
    crème fraîche
½ cup currant jelly

Equipment
1 high-sided pan, 10 inches
    in diameter

Arrange the bread pieces to cover the bottom of the pan, trimming the pieces to fit. Cover with the currants and add the cream mixture. Bake for about 30 minutes, making sure the pudding is cooked throughout. Remove from the oven and let cool for about 30 minutes.

Detach the edges with the tip of a knife. Invert onto a serving plate.

Melt the currant jelly over low heat. Serve in a sauce boat as an accompaniment. Garnish the pudding with currants.

Island Palette
(Palette des Îles)

This dessert plays with tastes and colors in an exotic way. I serve it chilled and stick a pineapple chip into each serving. If you cannot find ready-made chips, make them yourself—it is easy. Create a syrup by heating 1 scant cup sugar and 1 cup water; let boil for 1 minute. Cut thin slices of raw pineapple, dip them in the boiling syrup, and lay them immediately on a nonstick baking sheet. Bake at very low heat (120–140°F) for 2 hours until they are dried.

Photograph, opposite.

2 servings

Preparation: 30 minutes
No cooking required

1 mango
1 papaya
4 tablespoons butter,
    softened
²/₃ cup heavy cream, chilled
1 packet of vanilla sugar or
    2 teaspoons homemade
    (see page 7)
8 passion fruits
2 kiwis

Peel the mango and the papaya. Carefully cut the flesh into pieces, trying not to lose too much juice. Purée them separately in a food processor, adding 2 tablespoons butter to each fruit.

Put the cream into a deep medium bowl that is set into a larger bowl filled with ice water. Whip with an electric beater until the cream forms stiff peaks. Add the vanilla sugar while whipping.

Fold half of the whipped cream into each purée, carefully combining from bottom to top. Refrigerate.

Halve the passion fruits. Scoop out the flesh with its seeds and blend in the food processor. Strain through a sieve. Refrigerate.

Peel the kiwi and blend as well. Refrigerate.

With 2 soup spoons, shape plump scoops of the two mousses (mango and papaya), arranging two scoops of each type on each plate. Carefully pour some kiwi coulis and some passion fruit coulis around them without mixing the colors. Serve immediately.

COULIS, FRUIT DRINKS, AND SORBETS

Fruit takes on a new personality when transformed into a coulis, juice drink, or sorbet. In these forms the refreshing qualities of fruit are enhanced, as well as the color, taste, and fragrance.

Green, red, yellow, orange: colors are what pique your desire for something refreshing. I work with the colors of fruit the way a painter uses a palette. For example, if the color of a strawberry coulis is too light, I blend in a few raspberries, which will barely affect the taste but will make the coulis redder. When fruit is squeezed, the juice sometimes looks dull; sugar will brighten it. To save the fragile color of a pear from the damage that is quickly caused by oxidation, I immediately drizzle on some lemon juice. I accentuate the strength of the colors even more by creating contrast in the final presentation: I put a red fruit sorbet over a green or yellow coulis, or I serve two or three different colored sorbets in the same dish, topping them with fruit of yet another color.

A refreshment should also appear rich. I like it when a syrup or a sorbet looks like more than just flavored or chilled water. I like to see the bountiful, pulpy side of the fruit, an aspect that is often lost when the fruit is blended. In these cases, I often rely on sugar, which adds some warmth and texture. To give a red fruit coulis a lustrous allure without harming its fresh taste, I add a touch of raspberry or currant jelly.

I use ingredients with intense flavors to modify fragrance and taste. To make the bouquet of an apricot coulis stronger and more complex, I add two or three spoonfuls of blended apricot jam. The same kind of jam will counter the astringency of a passion fruit syrup. If I used a generous amount of sugar to brighten a sorbet, I balance the sweetness by adding a bit of bitter lemon. Lastly, instead of a still wine, I pour champagne over fruit for a more spirited, refreshing cocktail.

Island Fruit Drink.
Recipe page 133.

COULIS

These fruit sauces can be poured over ice cream, sorbets, fritters, charlottes, tarts, bread puddings, and more, giving them a splash of color and shine while enriching them with fresh taste and fragrance. Also use coulis to flavor the drinks you create. For example, my Moulin Cocktail is a simple combination of champagne and peach coulis.

So many uses for such an easy formula! To make a coulis, all you have to do is blend the flesh of the fruit, whether fresh, frozen, or in syrup. Sometimes this is all you need, but often this raw result is a little too sour. So, I sweeten it with confectioners' sugar or sugar syrup. On the other hand, I add zest to certain sugary coulis with a few drops of lemon. It is up to you to adjust the proportions to your liking. The proportions will vary almost every time, depending on the variety and maturity of the fruit.

For a thicker coulis, which might be desired to better coat a particular dessert, you might want to add gelatin to the mixture. Sometimes, I increase the amount of gelatin to compensate for the addition of a bit of cream liqueur. For example, I flavor a black currant coulis with *crème de framboise* (raspberry).

You will not know the final consistency of the coulis with gelatin until it is chilled. To save time, chill it in a bath of ice water, stirring it from time to time. You can always correct the coulis later. Adding sugar or gelatin to the coulis when it has returned to room temperature will not hurt it.

Very ripened apricots, peaches, and pears produce quite runny coulis. I get a thicker result by first cooking the fruit in a sugar syrup before blending. This way, I do not need to add much gelatin, if any.

You can use all the good tricks to improvise a coulis, adjusting its flavor, color, and texture according to your taste and what you have on hand. For example, I blend one part apricots in syrup with a one-quarter part apricot jam, then I thin it out with a nut-based liqueur (Amaretto or Noyau de Poissy) and the syrup from the can.

Pour your coulis immediately into a closed container when it's ready and use it soon—it will keep only twenty-four hours!

Red Fruit Coulis

(Coulis de Fruits Rouges)

Blending this coulis too long will lighten the color, producing a pink coulis instead of a beautiful bright red one.

Rinse the strawberries and remove the stems. Quickly rinse the raspberries and gently pat dry.

Place the fruit in a food processor with the water, lemon juice, and confectioners' sugar. Blend quickly. If necessary, add more sugar, water, or lemon juice. Strain through a sieve.

Variations

MANGO COULIS

Instead of using strawberries and raspberries, peel 2 ripe mangoes and cut the flesh into chunks, saving any juice. Proceed as with the red fruit coulis.

PEACH COULIS

Substitute peaches for the red berries. Peel 4 ripe peaches. Remove the pits and cut the flesh into pieces. Proceed as with the red fruit coulis.

1/2 pound strawberries
1/2 pound raspberries
1/3 cup water
2 tablespoons confectioners' sugar
A few drops lemon juice

Currant Coulis

(Coulis de Groseilles)

In a saucepan, stir the sugar and water over low heat until the sugar dissolves. Add the currants. Turn up the heat and whisk occasionally. When the sauce boils, remove from the heat.

Blend the mixture in the saucepan with an electric beater or in a food processor. Strain through a sieve, mashing the mixture with a spoon to extract the juice and pulp.

2 tablespoons sugar
1/3 cup water
1/2 pound currants

FRUIT DRINKS

Making a successful fruit drink entails knowing the fruit's characteristics. It is important to consider the fruit's density, which may vary greatly, for example, between a fleshy apricot, a juicy strawberry, and a very light currant. You cannot mix colors indiscriminately or your drink might turn into a brown or pale juice. You need to blend compatible flavors to achieve what wine enthusiasts call a bouquet. And you have to know how to crown the drink with a pretty garnish of flowers or fruit, chosen according to look or taste.

Finally, you cannot neglect the last detail that will make your drink a true luxury: dipping the edge of the glass in a green or red syrup poured onto a plate, then in sugar.

Fruit Juice
Cocktail
and Island
Fruit Drink
(recipes
page 133).

Apricot-Lime Fruit Drink
(Cocktail Abricot-Citron Vert)

Add your own personal touch to this drink. For example, instead of water substitute the same amount of white wine. Garnish the glass with a verbena branch and a slice of orange.

Combine all the ingredients in a food processor and blend well. Strain the mixture through a sieve, mashing it with a spoon to extract as much pulp and juice as possible. Serve immediately.

4 servings

1 pound ripe apricots, pitted
3/4 cup orange juice
Juice of 1 lime
Sugar, to taste
1/3 cup cold water
A few ice cubes

Strawberry-Apple Fruit Drink

(Cocktail de Fraises à la Pomme)

8 servings

2½ pounds strawberries
4¼ cups apple juice
Juice of 3 lemons
1 pinch nutmeg
A few ice cubes

Frost the edges of your glasses: dip them into grenadine syrup poured onto a plate, and then into sugar.

Stem the strawberries. Put them in a food processor along with the remaining ingredients and purée until smooth. Strain through a sieve, mashing with a spoon to extract the juice.

Pineapple-Banana Fruit Drink

(Cocktail Ananas-Banane)

2 servings

4 canned pineapple slices
1 banana
1 tablespoon honey
Juice of ½ lemon
1 ice cube

Packed with fruit, this drink can be enjoyed as a dessert or as a refreshing, vitamin-filled snack.

Place all the ingredients in a food processor. Purée until smooth. Serve immediately or refrigerate (no more than 1 or 2 hours).

Moulin Cocktail

(Coupe de Moulin)

Per glass
1 tablespoon cherry liqueur
1 fresh morello cherry
Champagne, chilled

Prepare this cocktail in advance by putting the glasses in the refrigerator and the champagne in a bucket. When the guests arrive, just pop the cork!

Put the cherry liqueur and a morello cherry in each champagne glass. Refrigerate. At serving time, fill the glasses with chilled champagne. Do not stir.

Vouvray and Champagne Cocktail

(Cocktail au Vouvray et au Champagne)

6 servings

Vouvray—white wine from Touraine that is dry, demi-sec, or sweet, still or sparkling—often tastes like green apple and quince, which makes it the perfect complement to fruit.

Decant the Vouvray. Wash the oranges and the lemon. Cut into rounds and stir into the Vouvray. Thinly cube the pineapple and slice the strawberries. Dust with sugar in a bowl. Separately refrigerate the fruit and the Vouvray for 1 to 2 hours.

    Strain the Vouvray through a sieve, mashing the citrus pulp with a spoon to extract the juice. Divide the wine–juice mixture among 6 frosted glasses. Drain the strawberries and pineapple and divide among the glasses. Fill the glasses with the chilled champagne. Do not stir.

2 oranges
1 lemon
2 cups still, dry Vouvray
1 cup champagne
2 slices fresh pineapple
1/2 pound strawberries
Sugar, to taste

Island Fruit Drink

(Cocktail des Îles)

6 servings

Photograph, pages 126 and 130.

Garnish each glass with fresh peppermint to enhance the taste of this exotic treat.

Peel and pit the mango; peel and seed the papaya. Cut into chunks and place in a food processor with the remaining ingredients, except the mint. Purée until smooth. Pour into glasses and garnish with the mint. Serve immediately.

1 mango
1 papaya
1 cup orange juice
1 cup pineapple juice
1 tablespoon lemon juice
Mint leaves, for garnish

Fruit Juice Cocktail

(Cocktail de Jus de Fruits)

2 servings

Photograph, page 130.

To garnish, arrange fruit—such as strawberries, lime wedges, and melon balls—on a wooden skewer, and lay it across the top of the glass.

Frost the edges of three glasses: dip them into the lemon juice and then into the sugar.

    Mix the fruit juices and the grenadine in a pitcher. Pour into glasses and add 2 ice cubes to each glass. Garnish as desired.

Juice of 1 grapefruit
Juice of 1 orange
Juice of 1 lime
2 tablespoons grenadine
Strawberries, melon balls, or lime wedges, for garnish

For the glass
Lemon juice, sugar

Le Peygros Sangria

(Sangria du Peygros)

6–8 servings

For the orange wine
3 bitter oranges
10 peppercorns
1 vanilla bean
½ cinnamon stick
4½ cups red or rosé wine
1 cup cognac
1 cup sugar

1 watermelon
Fruits of your choice
   (strawberries, oranges,
   pears, kiwi, melon, etc.)

Here is a refreshing watermelon surprise for a summer picnic with friends, which you will need to plan in advance: the orange wine needs to macerate for one month! I often serve this fruit-filled watermelon at "Le Peygros," my home in Mougins.

At least 1 month in advance: Quarter the oranges and place in a large jug. Add the peppercorns, vanilla bean, cinnamon stick, wine, and cognac. Let macerate for 20 days in a cool area away from light (but do not refrigerate). Add the sugar and continue to macerate another 10 days. Strain and reserve the orange wine.

The night before serving or the same day: Hollow out a watermelon, removing the seeds, to create a serving bowl. Fill with the orange wine. Refrigerate for 12 hours. Add your choice of peeled and sliced fruit. Serve in chilled glasses, if possible.

SORBETS AND GRANITAS

Lighter and more tart than ice cream, a sorbet contains no cream, eggs, or fat. It is made completely of puréed fruit, water, sugar, and sometimes lemon juice, which is used to awaken the fruit's flavor.

The recipes that follow are made from a base syrup that I make in advance. Either I poach the fruit in the syrup before blending, or I combine it with a juice or a purée of raw fruit. The next step is to place the mixture in an ice-cream freezer. One hour before serving move the sorbet to the refrigerator so that it will not be too hard.

Like a sorbet, a granita does not contain eggs, milk, or cream. It is made with sugar and fruit purée, fruit juice, and sometimes brandy. Instead of setting in an ice cream freezer, it is simply poured into a dish or metal bowl. (Do not use plastic, which does not conduct cold very well.)

When the granita has almost set, whisk it or run a spoon or the tines of a fork through it to break it up. This will prevent it from turning into a hard block. Repeat this process several times until it is completely set. The granita should have the texture of packed snow, dense but with distinct crystals. It is a refreshing sensation to bite into small pieces of flavored ice water, or to have them melt in your mouth. Granitas are served in frosted dishes.

Apricot Sauce

This key recipe will prepare apricots for many uses. The sauce can be used as is as a coulis, as the base for a sorbet, or as a marvelous syrup, to be served chilled, by adding 4 1/4 cups of mineral water and a touch of nut-based liqueur or kirsch. By replacing the mineral water with 1 bottle of champagne, you will create a cocktail. Serve it in a pitcher or a punch bowl, and garnish each glass with a verbena leaf.

INGREDIENTS
1 vanilla bean (split lengthwise)
1/3 cup wildflower honey
2 1/4 pounds apricots
1/2 cup sugar
juice of 1 lemon

In a saucepan, combine 1 cup water, the vanilla bean, and the honey. Bring to a gentle boil and add half the apricots, left whole. Stir to coat the apricots. Cover and simmer for 5 minutes. Remove from heat and let cool.

Halve and pit the remaining apricots. Combine in a bowl with the sugar and the lemon juice.

In a food processor, combine the honeyed apricots and the macerated apricots, along with their juice. Remove the vanilla bean from the syrup and add to the apricots. Blend into a thick purée. Strain through a sieve, mashing the mixture with a spoon to extract as much pulp and juice as possible.

Strawberry Sorbet (recipe page 138) shown here with a Mango Coulis (recipe page 129).

Base Syrup
(Sirop de Base)

2 1/2 pounds sugar
4 1/4 cups water

This syrup will keep 1 to 2 weeks in the refrigerator in a closed container. Use it as needed to make sorbets. In the summer, your supply will disappear quickly!

In a saucepan, whisk the sugar and water over low heat until the sugar has dissolved completely. Raise the heat and bring to a boil for 2 minutes. Remove from heat and let cool. Refrigerate in a closed container.

Lemon Sorbet
(Sorbet au Citron)

4 servings

1/2 cup lemon juice
3/4 cup base syrup, chilled
   (see above)
2/3 cup cold water

In a medium bowl, mix the lemon juice and the chilled syrup. Add the cold water and stir well.
   Pour the mixture into an ice-cream freezer. Follow freezer instructions until set.

Strawberry Sorbet
(Sorbet à la Fraise)

6 servings

1 pound strawberries
1 1/2 cups base syrup
1 tablespoon lemon juice
A few drops of vanilla
   extract or strawberry
   brandy

Wash, dry, and stem the strawberries. Purée in a food processor or blender. If you wish to remove the seeds, strain the purée through a fine sieve.
   Add the remaining ingredients to the purée and blend. Pour into an ice-cream freezer. Follow freezer instructions until set.

Variations

RED FRUIT SORBET
Proceed as with the strawberry sorbet, using the same proportions, but combine several kinds of fruit into the purée: strawberries, raspberries, wild strawberries, or red currants.

Red Fruit
Sorbet.

MULBERRY SORBET
Proceed as with the strawberry sorbet, using the same proportions, but omit the vanilla extract. Substitute mulberry brandy for the strawberry brandy.

Apricot Sorbet

(Sorbet à l'Abricot)

4–6 servings

You can make a very tasty sorbet, with touches of honey and vanilla, using my apricot sauce (see page 137). This is a simpler version that I also like.

In a saucepan, heat the syrup without bringing it to a boil. Add the apricots and poach over low heat for about 10 minutes. Remove from heat and let cool.

Purée in a food processor or blender, adding the liqueur or kirsch. Pour into an ice-cream freezer and follow freezer instructions until set.

1¾ cups base syrup (see page 138)
1 pound pitted apricots
1 tablespoon apricot liqueur or kirsch

Peach Sorbet

(Sorbet à la Pêche)

6 servings

Peel and pit all but one peach. Purée in a food processor or blender. You should have about 2 cups of pulp.

Add the syrup and the lemon juice and blend well. Pour into an ice-cream freezer and let set in the freezer. Slice the remaining peach into sections and use as a garnish.

2¼ pounds peaches
1⅓ cup base syrup (see page 138)
1 tablespoon lemon juice

The picture of refreshment. Clockwise, from top: Grapefruit Granita, Mint Granita, and Red and Black Currant Granita.

Red or Black Currant Granita

(Granité à la Groseille ou au Cassis)

4 servings

In a food processor or blender, blend the red currants with ½ cup sugar and the lemon juice. Strain through a sieve, mashing the mixture with a spoon to extract as much juice and pulp as possible. Pour into the mold of your choice and freeze.

As soon as the mixture begins to set, add 1 tablespoon of sugar and whisk. Return to the freezer. Repeat these steps 2 or 3 times until it has set completely.

½ pound red or black currants, weighed without stems
½ cup + 2–3 tablespoons sugar
A few drops lemon juice

Melon-Currant Granita

(Granité Melon-Groseille)

6 servings

*1 pound honeydew melon
pulp
¹/₂ pound currants
1¹/₂–2 cups sugar
Juice of ¹/₂ lemon*

In a food processor or blender blend the melon and currants with about 1¹/₂ cups sugar. Strain through a sieve, mashing the mixture with a spoon to extract as much pulp and juice as possible. Add the lemon juice and stir. Pour into a bowl and let set in the freezer.

Proceed with the same steps as with the currant granita (see page 141), gradually adding the remaining sugar.

Grapefruit Granita with Vermouth

(Granité au Pamplemousse et au Vermouth)

6–8 servings

*4¹/₂ cups grapefruit juice
¹/₃ cup + a few drops
vermouth
¹/₂ cup + 3 tablespoons
sugar*

In a blender, blend the grapefruit juice, ¹/₃ cup vermouth, and ¹/₂ cup sugar. Pour into a bowl and let set in the freezer.

As soon as the mixture begins to set, add 1 tablespoon of sugar and whisk. Return to the freezer. Repeat these steps 2 or 3 times until it has set completely.

At serving time, drizzle with a few drops vermouth.

Photograph, opposite.

Mint Granita

(Granité à la Menthe)

4–6 servings

*1 bunch mint
2 cups mineral water
1¹/₄ cups sugar
2 tablespoons mint liqueur*

I like this granita as an accompaniment for a fruit salad. It is also good with a red fruit coulis. This recipe will work with other herbs, such as rosemary or verbena; substitute a corresponding liqueur in place of the mint liqueur. Garnish with the leaves of the selected herb.

Photograph, page 140.

Coarsely chop the mint. Put in a bowl and set aside.

In a saucepan over low heat, stir the water and 1 cup of sugar until the sugar has dissolved. Raise the heat and bring to a boil for 2 minutes.

Pour the boiling syrup over the mint. Let cool. Strain through a sieve into a bowl. Add the liqueur and stir well. Place in the freezer.

Proceed with the same steps as in the currant granita recipe, gradually adding the remaining sugar.

PRESERVES
❦ AND ❦
CANDY

When I hear the word preserves, I think—nap time! As children, we would try in vain to take naps under a tree in the yard, our nostrils flaring each time the breeze carried the aroma from the kitchen our way. My mother and aunt would take advantage of the children's obligatory rest to cook jams, jellies, and fruit jelly candy in peace.

Once I took in the warm fragrance and closed my eyes, I could see the wonderful sight of cooking sugar: bursting bubbles, swirls of color changing with each passing minute. In my imagination, the preserving basin was like the crater of a volcano, a fiery world that would burst forth with incredible molten treats. To figure out when my nap time was over, I would wait until I distinguished a distinct caramel note.

It is wonderful to make your own preserves. Experience the pleasure of breathing in the aroma that each bubble releases when it bursts. Learn as I did during those hours of pretend sleep to recognize by smell when preserves are perfectly ready: the moment when you smell that the sugar is cooked enough and the fruit is still fresh.

The ritual of making preserves is one of the best ways to mark the passage from summer to winter. Seeing your jars lined up and your fruit jelly candy standing like many small rays of sun will warm you up. What a sorrowful place the world would be without the smell of preserves!

The four traditional red fruits: strawberries, raspberries, red currants, and cherries.
Recipe page 156.

JAMS AND JELLIES

There is nothing as simple as the principle of making jam. All you have to do is cook fruit with sugar. Setting depends on the amount of pectin, the substance naturally contained in fruit that is released while the fruit is cooking. Your preserves will jell if the pectin's environment is acidic enough. I often add lemon to my recipes to activate this process.

Many wonder what the difference is between jam and marmalade. It is more about the way each is made than their final taste. To make jam, fruit is cooked whole or in pieces in a sugar syrup, whereas to make marmalade, the fruit is left to macerate in sugar and release its juice before cooking. Jelly contains only sugar and fruit juice. There is no trace of pulp, and everything jells with pectin.

The Preserving Pan

The ideal pan in which to cook jams is a nonplated copper or nonreactive steel preserving pan that is washed well with vinegar and salt. The pan's rounded bottom is sup-posed to distribute heat better. Even though I feel nostalgia for this old-fashioned image, I know that today many kinds of pans will do just fine. In fact, the reason why preserving pans were used in the past is because standard saucepans, which often had thin bottoms, would cause the contents to stick. However, almost all of our modern saucepans have thick bottoms that distribute heat very well, so this is no longer a concern. Furthermore, the production of large amounts of jam required a large pot. Today, we have less time and generally make smaller quantities. Depending on the desired quantity of preserves, you can even use a pressure cooker without a cover, as long as the bottom is solid enough to prevent the jam from burning. The pan should be twice as large as the volume of sugar and fruit so that the contents do not overflow.

The Right Fruit for the Right Season

Choose shiny, firm fruit, without bruises from handling or from the beginning stages of rotting: this might introduce unwanted germs or give off a bad taste. Wash the fruit before cooking, being careful to just quickly rinse wild strawberries and raspberries.

A Few Ideas for Mixing Flavors

Red fruit:	*a tart touch, such as lemon or lime juice and zest, rhubarb*
Dark fruit:	*black currants, mulberries and brown sugar, cinnamon*
Peaches and nectarines:	*lemon zest, cloves, star anise, ginger*
Apricots:	*raspberry, honey, vanilla, almonds*
Mirabelle plums:	*brown sugar, lime*
Plums:	*cinnamon stick, crushed pepper*
Apples:	*peppercorns, nutmeg, cinnamon, lemon zest*
Pears:	*red fruit, lemon zest, ginger, vanilla, violet*
Figs:	*bay leaf*
Oranges:	*cinnamon*
Green tomatoes:	*thyme*

Always follow the calendar.

◆ In winter, until March: mandarins, oranges, grapefruits, lemons.
◆ In spring, until June: cherries, strawberries, rhubarb, apricots.
◆ In summer, until September: strawberries, raspberries, red currants, peaches, pears, melon, plums, grapes, mulberries, blueberries, black currants.
◆ In September: plums, blueberries, mulberries, figs, grapes.
◆ In fall, until December: quince, chestnuts, pears, apples.

Extra Flavors

I like to add unusual or exotic touches to my preserves.

◆ Vanilla, cinnamon, ginger, star anise, candied ginger, cardamom, nutmeg, cloves, pepper, or allspice can be added at the start of cooking.
◆ Walnuts, almonds, and hazelnuts should be added a few minutes before you are done cooking.
◆ Mint, rose, basil, thyme, or bay leaf is added at the beginning. The leaves can be left whole or chopped and

The Amount of Sugar

Sugar is what ensures that your preserves will keep. Therefore, you need enough sugar—but not too much—to guarantee both longevity and balanced taste. In general, we use equal amounts of sugar and fruit pulp, once peeled. You can increase the proportion of sugar for watery fruit, and decrease it for fruits rich in pectin (lemons, quince, currants, mulberries, oranges, apples) or if you are adding pectin. If there is not enough sugar, or if it has not cooked long enough, your preserves run the risk of fermenting. If there is too much sugar, the preserves will crystallize.

I prefer to use granulated sugar, but I sometimes substitute honey for a portion of the sugar. Try it with currants and raspberries! Sugar sold specifically for preserves is mixed with pectin and citric acid. These elements will accelerate the setting and therefore shorten the time needed for cooking. I do not use this product, preferring to calculate my own amount of pectin and cooking time to achieve better taste.

wrapped in a cheesecloth bundle to be removed later.

◆ Jasmine flowers, acacia, dandelions, marigolds, and violets are nice to look at and smell in jam. Remove the base of the petals with scissors: this part is bitter. If you are using dried petals, scald them for 1 minute and drizzle on just a bit of lemon juice.

◆ Rum, kirsch, muscat wine, and orange liqueur are added just before canning.

The Test: Are Your Preserves Ready?

When you start cooking your preserves, the water from the fruit evaporates and a lot of steam rises. Once evaporation slows, remove the froth that appears around the edges of the surface with a skimmer. Your preserves will be nice and clear in the end.

The liquid will then become more dense, which is the time to test it. Drop a bit of jelly on a plate. It should congeal as it cools, forming a drop that is not very runny. At that point, remove immediately from the heat. The preserves are ready to be canned.

Canning

Prepare the jars beforehand, as well as their lids. Wash them with very hot water and dry them with a clean towel, or turn them over on a towel and let them dry.

My dear mother would spend a long time cutting out paper circles and soaking them in alcohol before resting them on top of the preserves. Then she would close the jar with cellophane and a string. This way she would fight against mold.

Airtight jars, the kind with screw-on tops or those used to make commercial, vacuum-packed preserves are a lot less work. Fill the jar with boiling fruit but do not let it overflow. Close the jar immediately and flip it over for 5 minutes, then turn it right side up. The small amount of air in the jar will be pasteurized once it makes contact with the heat. As they cool, the preserves will decrease in volume, creating a sufficient vacuum to guarantee a proper seal.

There is another canning method if you are uncertain about the quality of the lids.

Allow the preserves to cool and then place a sheet of cellophane over the jar before closing.

The sides of the jar should not be stained with preserves. To fill the jar properly, use a special wide-neck funnel that you can find in kitchenware stores.

Your creations will keep for several months in a dry, dark place or in the refrigerator. In my recipes, I do not indicate how many jars you will need because the number will vary according to their shape and size. Big or small? Consider how fast those around you will eat the preserves, keeping in mind that an open jar will keep only a few days.

Reduced-Sugar Preserves

Preserves that contain less sugar will not keep as well as those with full-strength sugar, unless you use my little trick. Make your preserves using only 60–75% of the amount of sugar. Pour them in jars with rubber closings (mason jars) made for preserving. Let cool, close, and then seal in water for 30 minutes at 195°F. Once the jar is opened, store in the refrigerator and eat quickly.

Apricot-Peach Jam with Vanilla

(Confiture d'Abricots et de Pêches à la Vanille)

2¹/₂ pounds apricots
2¹/₂ pounds peaches
4¹/₂ pounds sugar
Juice of 1 lemon
3 vanilla beans, split
*   lengthwise*
2 handfuls fresh almonds

Wash the fruit. Halve and pit the apricots.

In a large saucepan of boiling water, scald the peaches for 1 minute and drain. Peel, pit, and cut into pieces.

Combine 1 cup water, the sugar, and the lemon juice in the saucepan. Bring to a boil for 4–5 minutes. Add the peaches, apricots, and vanilla beans. Cook for approximately 20 minutes, stirring regularly.

Test (see page 150). Add the almonds. Skim the surface. Remove the vanilla beans and cut into pieces. Pour the preserves into canning jars, adding a piece of vanilla bean to each jar. Seal immediately (see page 151).

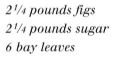

Fig Jam with Bay Leaf

(Confiture de Figues au Laurier)

2¹/₄ pounds figs
2¹/₄ pounds sugar
6 bay leaves

Wash, dry, and cut the figs into pieces.

In a large saucepan, bring 1 cup of water and the sugar to a soft boil (235–240°F). If you do not have a thermometer, drip some boiling syrup into a bowl of ice water and then take it between your fingers: you should be able to form a small, soft ball.

Add the figs and cook for 35–40 minutes.

Test. Add the bay leaves. Skim and pour into canning jars, putting one leaf in each jar. Seal immediately. (See pages 150–151.)

Photograph, opposite.

Quince Jam
(Confiture de Coings)

2¼ pounds quince, skinned
Juice of 1 lemon
2¼ pounds sugar

Choose ripe quince. Slice them into sections and stir immediately into cold water mixed with the lemon juice.

In a preserving pan or large saucepan, boil 2 cups water and the sugar. Add the quince and cook for about 1 hour over low heat, stirring occasionally.

Test, skim, and pour into canning jars. Seal immediately. (See pages 150–151.)

Melon-Raspberry Jam with Star Anise
(Confiture de Melon et de Framboise à la Badiane)

1 pound melon flesh,
    without seeds
4 cups sugar
Juice of 1 lemon
4 star anise
1 pound raspberries

Scoop out the melon flesh and cut into cubes. In a large saucepan or preserving pan, combine 1 cup water with the melon, sugar, lemon juice, and star anise. Bring to a boil and cook for about 30 minutes, stirring regularly.

Add the raspberries and cook for 15 minutes more, stirring regularly.

Test, skim, and pour into canning jars (see pages 150–151). Divide the star anise among the jars (if necessary, break into pieces). Seal.

Photograph, page 157.

My Friend Guy Gedda's Green Tomato Jam

(Confiture de Tomates Vertes de mon Ami Guy Gedda)

Wash the tomatoes and remove their stems. Cut into sections the size of orange slices.

In a nonreactive steel container, arrange a layer of tomatoes and dust with sugar. Repeat until you use all the tomatoes and sugar. Refrigerate and let macerate for 20–24 hours, stirring 2 or 3 times with a wooden spoon.

Pour the tomatoes into a preserving pan or a nonreactive saucepan.

Cut the lemons (do not peel) into ⅛-inch slices. Cut each slice in half.

Bring tomatoes to a boil and reduce heat to low. Skim. Cook for one hour. Add the lemons and cook for 1 hour more.

Test, skim, and pour into canning jars while still hot. Seal immediately. (See pages 150–151.)

4½ pounds green tomatoes
5 cups sugar
3 lemons

Grape Jam

(Confiture de Raisins)

Wash and stem the grapes. In a large saucepan or preserving pan, combine the grapes and 1 cup water. Cook for 15 minutes over low heat, stirring occasionally. With a skimmer, remove the seeds that rise to the surface.

Add the sugar and cook for 35–40 minutes.

Test, skim, and pour into canning jars. Seal immediately. (See pages 150–151.)

2¼ pounds black grapes
3¾ cups sugar

Jam of Four Red Fruits

(Confiture aux Quatre Fruits Rouges)

4 1/2 pounds sugar

1 pound morello (sour)
   cherries

2 vanilla beans, split
   lengthwise

Juice of 1 lemon

1 pound strawberries

1 pound raspberries

1 pound red currants

Wash and stem the fruit. Pit the cherries.

In a large saucepan or preserving basin, combine the sugar and 3 cups water. Cook without stirring until syrup comes to a soft boil (235–240°F). If you do not have a thermometer, drip some boiling syrup into a bowl of ice water and then take it between your fingers: you should be able to form a small, soft ball.

Add the cherries, lemon juice, and vanilla. Cook for 10 minutes, stirring occasionally. Add the strawberries and cook for 5 minutes. Add the raspberries and red currants and cook for 5 minutes more.

Test, skim, and pour the boiling jam into canning jars. Seal. (See pages 150–151.)

Apricot Marmalade with Their
Sweet Almonds and Vanilla

(Marmelade d'Abricots aux Amandes Douces et à la Vanille)

2 vanilla beans

2 1/2 pounds apricots

2 1/4 pounds sugar

Juice of 1 lemon

Split the vanilla beans lengthwise and cut into 1-inch sticks.

Halve and pit the apricots, reserving the pits. In a large saucepan or preserving pan, combine the apricots, sugar, lemon juice, and vanilla beans. Mix well and let macerate for 4–5 hours at room temperature, stirring occasionally.

In the meantime, crack the pits with a nutcracker and extract the almonds. Peel.

Bring the apricot mixture to a boil over high heat, stirring often. Cook 30 minutes, stirring often and scraping the bottom.

Test and skim (see page 150). Remove the vanilla beans.

Add the almonds from the pits and stir. Pour the boiling jam into canning jars, dividing the vanilla pieces among the jars. Seal immediately (see page 151).

Jam of Four Red Fruits (top). Melon-Raspberry Jam with Star Anise (bottom); recipe page 154.

4 Fruits
aout 97

Rhubarb Marmalade

(Marmelade de Rhubarbe)

Photograph, opposite.

Wash, peel, and cut the rhubarb stalks into pieces. Place them in a container with the sugar. Let macerate for 10–12 hours.

Drain the pieces and pour the juice into a preserving pan or large saucepan. Bring to a boil. Add the rhubarb and cook for 20 minutes, stirring regularly.

Test, skim, and pour into canning jars. Seal. (See pages 150–151.)

4 1/2 pounds rhubarb stalks
4 1/2 pounds sugar

Papaya Marmalade with Spices

(Marmelade de Papayes aux Épices)

Peel, seed, and cut the papaya into pieces. Combine with the sugar and the 5 spices in a large bowl. Cover and let macerate for 2 hours.

In a preserving pan or large saucepan combine the papaya mixture with 1 cup water and the lemon juice. Cook for 1 hour over low heat, stirring occasionally.

Test, skim, and pour into canning jars. Seal. (See pages 150–151.)

2 1/4 pounds papaya,
weighed peeled and
seeded
2 1/4 pounds sugar
1 pinch nutmeg
1 pinch cloves
1 pinch grated ginger
1 star anise
1 vanilla bean
A few drops lemon juice

Plum Marmalade with Lavender Honey

(Marmelade de Reines-Claudes au Miel de Lavande)

Pit the plums. Combine in a large bowl with the sugar and lemon juice. Let macerate for 6 to 8 hours at room temperature.

In a preserving pan or large saucepan, combine the plum mixture and the honey. If you have some fresh lavender, add it. Bring to a boil and cook for approximately 45 minutes.

Test, skim, and pour into canning jars. Seal. (See pages 150–151.)

2 1/2 pounds plums, such as
Reine Claudes
2 1/4 cups sugar
Juice of 1 lemon
1 1/2 cups lavender honey
Fresh lavender (optional)

Mango Jam with Ginger

(Confiture de Mangues au Gingembre)

Photograph, opposite.

Peel the lemon zest over a preserving pan or large saucepan. Peel and grate the ginger and add to zest. Add sugar and 1 cup water to the pan and mix well. Cook for 10 minutes, stirring occasionally.

In the meantime, peel, pit, and dice the mangoes. Juice the lemon.

Add the vanilla bean, mangoes, and lemon juice to the pan. Continue to cook, stirring occasionally, for 20 minutes.

Test, skim, and pour into canning jars. Seal. (See pages 150–151.)

Zest and juice of 1 lemon
1 bulb ginger, peeled and grated
2 1/4 pounds sugar
2 1/4 pounds mangoes, weighed peeled and pitted
1 vanilla bean, split lengthwise

Traditional Red Fruit Jelly

(Gelée de Fruits Rouges Tradition)

This classic recipe will work with strawberries, raspberries, mulberries, and blueberries. Apples, quince, and lemons can also be prepared in this way.

In a blender or food processor, purée the fruit. Strain through a fine sieve, reserving the juice.

In a saucepan over medium heat, dissolve the sugar in 1/3 cup of water. Add the fruit juice and lemon juice.

Cook over medium heat for approximately 30 minutes. Test, skim, and pour into canning jars. Seal. (See pages 150–151.)

Fruit to yield 4 1/4 cups strained fruit juice
4 1/4 cups sugar
Juice of 1 lemon

Quick Raspberry Jelly

(Gelée de Framboise Rapide)

In a blender or food processor, purée the fruit. Strain through a fine sieve, reserving the pulp and juice. Place in a saucepan over medium-high heat.

In a small saucepan, combine the pectin with the same amount of sugar. When it comes to a boil, add this mixture to the raspberries. Boil for 2 minutes, stirring constantly. Add the remaining sugar and cook for 2 minutes more.

Test, skim, and pour into canning jars. Seal. (See pages 150–151.)

2 1/4 pounds raspberries
3 tablespoons + 1 teaspoon pectin
2 1/4 pounds sugar

Red or Black Currant Jelly

(Gelée de Groseilles ou de Cassis)

2¼ pounds red or black
currants
Approximately 2¼ pounds
sugar

Wash, dry, and stem the currants. In a preserving pan or large saucepan over low heat, combine the currants with 3 tablespoons water. Do not boil. Stir for 10 minutes with a wooden spatula until the currants burst.

Set a colander lined with cheesecloth over a bowl. Pour the currants into the cheesecloth and make a bundle, bringing together the corners of the cloth at the top. Twist the cloth to tighten the bundle and squeeze well with your hand. (Wear gloves: juice stains!)

Kitchenware stores offer a very simple tool made of wood and nylon netting that will facilitate the extraction process. You can also run the fruit through a food mill that squeezes and purées, as long as you do not crush the seeds too much.

Measure the juice and pour it into the preserving basin or saucepan. Add the same amount of sugar. Cook for 25–30 minutes over medium heat.

Test, skim, and pour into canning jars. Seal immediately. (See pages 150–151.)

Red Currant
Jelly.

Quince Jelly

(Gelée de Coing)

2¼ pounds quince, weighed
peeled and seeded
2¼ pounds sugar
Juice of ½ lemon

Peel, seed, and quarter the quince. Place in a preserving pan or large saucepan with water to cover. Cook over low heat until fruit is tender.

Strain through a colander and measure the reserved juice: you need 4¼ cups. Return liquid to the pan and combine with the sugar and lemon. Cook for 30 minutes over medium heat.

Test, skim, and pour into canning jars. Seal immediately. (See pages 150–151.)

Lemon Jelly with Verbena Leaves

(Gelée de Citron aux Feuilles de Verveine)

1 branch verbena
4¼ cups lemon juice
Zest of ½ lemon, grated
¼ cup pectin
2½ pounds sugar

Remove the verbena leaves from their stems. In a saucepan, heat the lemon juice and zest with the verbena leaves.

Combine the pectin with ¼ cup sugar and add to the lemon juice. Boil for 2 minutes, stirring constantly. Add the remaining sugar and let boil for 2 minutes more.

Test, skim, and pour into canning jars. Seal. (See pages 150–151.)

Bitter Orange Jelly

(Gelée d'Orange Amère)

8 bitter oranges
2 sweet oranges
1 lemon
Sugar

This recipe was given to me by my friend Florence, heiress of the great Polvérino family of restaurant owners. She has shared her talent with the big names of our beautiful village of Mougins—Pablo Picasso, Francis Picabia, Christian Dior—and she continues to provide us with joy today.

Cut the fruit into small pieces. Do not peel.

Remove the seeds and place in a bowl with water to cover. Let soak for 24 hours.

Place the fruit in a preserving pan or large saucepan and cover with water (about ½ inch above the fruit). Let soak for 24 hours.

Heat the pan until the peel of the fruit softens. (It should be supple in your fingers, without completely falling apart.)

Measure the contents of the pan. Return to the pan and combine with an equal amount of sugar.

Strain the seeds, reserving the "jelly." Add jelly to the warm fruit mixture. Tie the seeds in a cheesecloth bundle and bury into the fruit. Cook for about 30 minutes until the jelly sets. Remove the cheesecloth bundle.

Test, skim, and pour into canning jars. Seal while hot. (See pages 150–151.)

CANDY

Making candy from fruit is a terrific way not only to preserve fruit, but to create elegant decorative flourishes. Key ingredients such as sugar, almond paste, and chocolate transform fruit into wonderful confections. Whether raw or cooked, fruit candies are as delicious as they are beautiful.

Dressing Up Fruit

It is easy to dress up fruits by dipping them in chocolate, coating them in sugar, or combining them with almond paste that is lightly tinted a pastel pink or green. Candy-coated fruit brings a festive touch to your dessert presentations. Offer your guests one last temptation while having coffee or after-dinner drinks. I like to eat these treats with a small glass of Cognac, a muscat wine from Frontignan, or Grand Marnier for the luscious combination of fruity fragrances and sweet, robust flavors.

Do not prepare the fruit candies too far in advance; they are much softer and fresher in the first few days. To store them for a few days, keep them in small paper boxes in a dry place.

Cherries

Halve large, candied cherries. Mix a drop of kirsch into almond paste and color it a light pink with red food dye. Wedge a small ball of paste between the cherry halves and roll in your palm until you have an even sphere with a light band across the middle. Roll in a plate of crystallized sugar.

Walnuts

Flavor some almond paste with a bit of walnut liqueur and very concentrated coffee. Wedge a small ball of paste between 2 shelled walnut halves and squeeze together. Dip briefly in beaten egg white and roll in a plate of crystallized sugar.

Prunes

Pit the prunes. Fill the inside with almond paste that has been flavored with Armagnac and tinted with a drop of red food coloring. Roll in crystallized sugar.

Dates

Pit the dates. Fill the inside with a small ball of almond paste flavored with Maraschino and tinted a pale green with a drop of food coloring. Roll in crystallized sugar.

Almonds

Tint some almond paste a very light green with food coloring. Make small oval nuggets with the paste. Place these between 2 whole, blanched almonds. Dip briefly in beaten egg white and roll in crystallized sugar.

Candied rinds of small bitter oranges.

Strawberries

Place squares of dark chocolate (60% or more cocoa) in the top of a double boiler. As soon as the chocolate is soft, remove from the heat and stir until smooth with a wooden spoon. Let rest for a few minutes. Tilt the saucepan. Dip large strawberries that have been washed and dried into the melted chocolate, holding them by the stem.

Small Bitter Oranges

Buy ready-to-eat candied orange or lemon rinds at a bakery or candy shop. Slice into sticks measuring about ¼ inch in width. Melt a bar of dark chocolate (as with the strawberries, above). Let rest. Dip the rinds 3 at a time in the warm chocolate. Remove them with a trussing needle or a fork. Let harden on a rack.

Kumquats in Multicolored Chocolate

(Kumquats aux Chocolats Multicolores)

4 servings

These tiny citrus fruits from the Far East with a slightly bittersweet taste are eaten raw or candied, and always with their rinds. They work just as well in a dessert as in a savory dish. Personally, I prefer them as a delicate treat with coffee or tea. Here is a very colorful way to prepare them as an accompaniment.

Wash and dry the kumquats.

With a peeler, grate 1 ounce of white chocolate and 1 ounce of dark chocolate onto a plate. Mash the shavings with a fork (not with your fingers, or you will melt them).

In separate double boilers or in separate small pans set into a larger pan of water, heat the remaining dark chocolate, the remaining white chocolate, and the fondant. The temperature of the fondant should not exceed 65°F.

Dip ¾ of each fruit into the fondant, holding them by the stem or sticking the end with a toothpick. Lay them on a sheet of aluminum foil. Let dry.

Tint the remaining fondant with a few drops of red food coloring to achieve a soft pink.

Scatter the chocolate sprinkles and colored sprinkles on 2 plates.

Remove the chocolates and the fondant from the heat.

Dip 4 kumquats in the dark chocolate and lay them on the crushed chocolate. Dip 4 kumquats in the white chocolate and lay them on the chocolate sprinkles. Dip the 4 remaining kumquats in the pink fondant and lay them on the colored sprinkles.

Allow to harden on wax paper. Store in small boxes. Present on a pretty plate with a white doily and decorate with a few flowers.

12 kumquats

3½ ounces + 1 ounce bitter
  dark chocolate

3½ ounces + 1 ounce white
  chocolate

5 ounces white chocolate
  fondant (from a pastry
  shop)

4 tablespoons dark
  chocolate sprinkles

4 tablespoons colored
  sprinkles

A few drops red food
  coloring

Fruit Jelly Candy

(Pâtes de Fruits)

These candies are like hardened jelly that you can eat with your fingers. Made from a pulp of fruit and sugar, they will be rich in natural pectin. Refer to the chart below for proportions. These treats fit right in with winter holiday goodies.

Select ripened fruit. Wash, peel, stem, and pit or seed. In a food processor or blender, blend the pulp into a fine purée.

In a nonreactive saucepan, combine the purée with $9/10$ of the sugar. Bring to a boil, whisking constantly.

In a bowl, combine the remaining sugar and the pectin. Add to the saucepan and continue to cook, whisking constantly, until the mixture reaches 220–230°F. If you do not have a thermometer, put a bit of the jellied sauce on a spoon. Let it cool slightly, and take some between your fingers. When you separate your fingers, a thin strand of jelly, less than $1/8$-inch thick, should break as you widen the space between your fingers. This is called the "thread stage."

Pour the mixture immediately onto parchment paper. Surround the surface with metal rulers or place the paper on a tray with raised sides. The paste layer should be about $1/3$ inch thick.

Let cool completely. Cut out candies in the form of your choice with a cutter or a thick blade.

Finally, roll the pieces in sugar and let dry for 2–3 hours. Store in a sealed container, separating the layers with wax paper.

An assortment of fruit confections: kumquats with chocolate, jellied fruit candy, and nuts and cherries with almond paste.

Ingredients and proportions

FRUIT	PULP	SUGAR	PECTIN	COOKING TIME (once boiling)
apricots	2¼ pounds	2⅓ pounds	5 tablespoons	8 minutes
citrus fruit	1 cup whole fruit blended with 3 cups juice	2⅓ pounds	5 tbls + 1 tsp	8 minutes
black currants	2¼ pounds	2⅓ pounds	4 tablespoons	8 minutes
cherries	2¼ pounds	2¼ pounds	4 tbls + 2 tsps	8 minutes
quince	2¼ pounds	2⅓ pounds	3 tbls + 1 tsp	5 minutes
red fruit	2¼ pounds	2¼ pounds	4 tbls + 2 tsps	5–8 minutes
red currants	2¼ pounds	2⅓ pounds	4 tablespoons	8 minutes
blueberries	2¼ pounds	2⅓ pounds	5 tablespoons	8 minutes
peaches	2¼ pounds	2⅓ pounds	5 tablespoons	10 minutes
plums	2¼ pounds	2¼ pounds	5 tablespoons	12 minutes

Crystallized Flowers and Leaves
(Fleurs et Feuilles Cristallisées)

Flowers and/or leaves
1/3 cup gum arabic
Sugar

You can decorate your desserts with all kinds of flowers and leaves: bunches of tree flowers, such as pear, cherry, or apple blossoms; pansies, small roses, or sprigs of lavender; as well as black currant leaves, violet or mint leaves, and verbena or jasmine leaves. My trick is to add a few drops of rose water or orange blossom water into the soaking liquid, which will enhance the taste of these edible flowers.

Boil 2 cups water and dissolve the gum arabic. Strain and then let rest for 12 hours at room temperature.

Dip the leaves or flowers in the gum arabic and bury immediately in the sugar. Let rest for about 20 minutes. Remove and shake lightly to loosen excess sugar.

Place the leaves or flowers in a sealed container, the bottom of which should be covered with sugar. They will keep up to 3 weeks.

It will also work to dip the flowers in beaten egg whites mixed with very little water before covering them with sugar. They will keep only one day using that method.

Crystallized pansies.

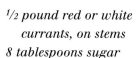

Crystallized Currants
(Groseilles Cristallisées)

1/2 pound red or white
currants, on stems
8 tablespoons sugar

These currants are delicious, pretty, and easy to make. They do not last more than 48 hours.

Refrigerate 1/2 pound of red or white currant bunches for 2 hours. When removed, they should be lightly and evenly damp from the condensation.

Sprinkle the sugar on a tray. Roll the currant bunches in the sugar to coat completely. Set aside in a dry place for 1 hour.

CONSERVES

Here are a few recipes for homemade conserves that seal up summer's kindness for the winter's pantry. There is one simple rule for all conserves: use perfect, healthy fruit and perfectly clean jars. You need to stick holes in the fruit before sealing the jars so that they do not burst under the heat. This will also allow the pulp and the brandy syrup to intermingle: the fruit will sweeten or soak up brandy, and the syrup will take on the rich flavor of the fruit.

Purple Figs with Bay Leaf

(Figues Violettes au Laurier)

2¹/₄ pounds purple figs
1¹/₄ cups sugar
A few bay leaves

Pierce the figs in 3 or 4 places with a thick trussing needle (or a small knitting needle). Place them in sterilized jars, leaving 1¹/₂ inches at the top (choose your jars according to the size of the figs).

Dissolve the sugar in 2 cups cold water. Pour this syrup into the jars, leaving ¹/₂ inch at the top. Add 1 bay leaf to each jar.

Seal in a boiling-water bath for 40 minutes and let cool on a rack.

Apricots with Vanilla or Verbena

(Abricots à la Vanille ou à la Verveine)

Pierce the apricots in 3 or 4 places with a thick trussing needle (or small knitting needle). Place them in sterilized jars, leaving 1½ inches at the top (choose your jars depending on the size of the fruit).

Dissolve the sugar in 2 cups cold water. Pour this syrup into the jars, leaving ½ inch at the top. Add 1 vanilla bean or a few verbena leaves to each jar.

Seal in a boiling-water bath for 40 minutes and let cool on a rack.

2–4 pounds apricots
1¼ cups sugar
Vanilla beans or a few verbena leaves

Mirabelle or Reine Claude Plums in Brandy

(Mirabelles ou Reines-Claudes à l'Eau-de-Vie)

Wash, stem, and dry the plums. Place in a jar without cramming them. Fill the jar with enough brandy to cover the plums. Let rest 2 weeks.

Add 1½ cups superfine sugar. Close the jar tightly. The plums will be good to eat in 5–6 weeks.

2¼ pounds plums, such as Mirabelle or Reine Claude
Approximately 2 cups brandy
1½ cups superfine sugar

Cherries in Brandy

(Cerises à l'Eau-de-Vie)

Proceed as with the plums (see above). Cut the stems halfway. Do not include sugar.

Choose tart cherries: the Montmorency variety, which are also known as "English" cherries, or morellos. I prefer them to be a bit firm with crunchy flesh so that they do not soften too much over time.

LITTLE CHEFS
IN THE KITCHEN

Whatever you do, do not keep your children away from the mysteries of the kitchen. They bring a freedom and spontaneity to cooking that adults have often lost. Children love to taste things as they go along, make crazy shapes and decorations, and devour everything in sight without worrying about their weight.

Allowing children to make fruit desserts is a good way to cultivate their senses. You will have the pleasure of hearing their satisfaction—almost every child has a sweet tooth. They are the best audience and judges, never holding back their ideas, and they honor their desserts (and yours) with total sincerity. Shamelessly, they let their joy burst forth.

So save them a place at the counter, even if they are bothering you a bit, and offer them a bench to stand on so they can see better, smell better, and cook better. Just make sure they are safe and do not touch certain utensils.

The following recipes are adapted to children's taste, and often they involve fun presentations. Children can help you make these desserts, and some of them they will be able to make on their own. Each recipe will show them that creating a successful dessert that suits their palate is not so difficult or complicated.

Apricots and Cream

(Abricots au Plat)

4 servings

Preparation: 20 minutes
Cooking: 3–4 minutes

8 apricot halves, in syrup
1 tablespoon cornstarch
1 cup milk
¼ cup sugar
1 pinch salt

Equipment
4 white porcelain dishes

This recipe will be even better if you use homemade vanilla sugar (see page 7).

Drain the apricot halves, reserving the syrup.

Place the cornstarch in a bowl. Gradually dilute with 2 tablespoons cold water.

In a saucepan, heat the milk, sugar, and salt. When it comes to a boil, add the cornstarch, whisking constantly for 30 seconds to thicken the cream. Remove from heat and divide among 4 dishes. Place 2 apricot halves in each dish, rounded side facing up. Let cool.

Drizzle with the reserved syrup. Serve immediately.

Chocolate-Coated Pear and Vanilla Ice Cream

(Poire au Chocolat et Boule de Vanille)

4 servings

Preparation: 10 minutes
Cooking: 2–6 minutes

8 pear halves, in syrup
1 bar dark chocolate (at least 3½ ounces)
¾ cup milk
⅓ cup heavy cream
4 scoops vanilla ice cream

This delicious dessert creates an exciting hot-and-cold sensation in your mouth.

Drain the pears. Set aside.

Break the chocolate into bits and place in the top of a double boiler over low heat. When the chocolate begins to melt add the milk and cream, stirring regularly until the chocolate melts completely.

Divide the pear halves among 4 plates. Pour the chocolate cream on top. Add a scoop of vanilla ice cream to each plate and serve.

Rolled Cake
(Biscuit Roulé)

6 servings

Choose a jam with a good deal of whole fruit or fruit pieces. The sweet pulp will contrast in taste and texture with the lightness of the cake. To sift the flour, cornstarch, and yeast together, place all three ingredients in a fine sieve held over a bowl and stir the mixture with a spoon. Children enjoy seeing the pretty white powder fall like snow!

Preheat the oven to 400°F. Lay parchment paper on the jelly roll pan and grease with the butter.

In a medium bowl, beat the egg yolks, half the sugar, and the vanilla sugar until you have a light-colored, frothy cream.

In another bowl, whip the egg whites and 1 pinch salt with an electric beater to form stiff peaks. Add the remaining sugar, continuing to whip. The peaks will become a bit thicker.

Fold the egg whites into the egg cream, carefully stirring from bottom to top.

Sift together the flour, cornstarch, and baking powder. Carefully fold into the egg mixture until well blended.

Pour the batter onto the baking sheet and smooth the surface with a spatula.

Bake for 10 to 15 minutes.

Powder a slightly damp dish towel with sugar.

Remove the cake from the oven. Using the parchment paper, carefully flip the cake over onto the towel. Remove the paper and roll up the cake in the towel, rolling from one long side to the other. Let cool.

Unroll the cake and spread the jam on top. Roll it up again without the towel. The seam should be on the bottom.

Refrigerate. Slice to serve.

Preparation:
30–40 minutes
Cooking: 10–15 minutes

2 tablespoons butter
4 eggs, separated
½ cup sugar
2 teaspoons vanilla sugar (see page 7)
1 pinch salt
½ cup flour
½ cup cornstarch
1 pinch baking powder
Your choice of fruit jam

Equipment
Jelly roll pan, 15½ × 10½ inches
Parchment paper
Dish towel

Hodgepodge of Red Fruit
on an Almond Crust

(Méli-Mélo de Fruits Rouges sur Croûte d'Amandes)

4 servings

Preparation: 30 minutes
Cooking: 25 + 10 minutes

For the almond cream pastry
5 tablespoons butter,
    softened + more to
    grease pans
1/3 cup sugar
1/3 cup ground blanched
    almonds, pulverized in
    a food processor
1 teaspoon cornstarch
2 eggs

For the coulis
1/4 pound raspberries
1/2 pound strawberries
1 1/4 cups sugar

For the topping
1/4 pound raspberries
3/4 pound strawberries

Apricot topping
    (see page 137)
Mint leaves, for garnish

Equipment
2 tartlet pans, 4 inches
    in diameter

I make these flavorful tartlets with strawberries that grow on the outskirts of Nice—we think our strawberries are the best! Add a final touch by decorating each tart with a sprig of mint. You can also make one single tart in a high-sided pan, 8 1/2 inches in diameter.

Preheat the oven to 350°F.

Prepare the almond cream pastry. In a medium bowl, whip the softened butter and 1/3 cup sugar for 1 minute. Add the ground almonds and cornstarch, and beat for 1 minute. Add the eggs one at a time, beating each vigorously for 1 minute.

Grease the tart pans with butter. Pour in the cream and distribute evenly. Bake for 20–25 minutes. Let cool. Remove from the pans.

Prepare the coulis. Purée the raspberries and strawberries in a blender or food processor. Add 1/2 cup sugar and blend again.

Strain the coulis through a sieve, reserving the raspberry and strawberry seeds. Place them in a saucepan with 3/4 cup sugar. Cook for approximately 10 minutes over low heat, stirring with a wooden spoon. Add a little water, if needed.

Spread a thin layer of the seed jam on the crusts and in each mound the raspberries and strawberries. Cover the bottom of each serving plate with the coulis. Cover this layer with apricot topping. Place a tartlet on one side of each plate. Garnish with mint leaves.

Pineapple Pine Tree

(Sapin d'Ananas)

4 servings

This playful dessert is as fun to make as it is good to eat.

Cut the top of the pineapple and save the crown. Wash, dry, and separate the leaves.

Quarter the pineapple lengthwise. Cut away the flesh along the inside of the rind to separate. Cut the flesh into small slices and slide out every other slice so that it extends over the rind.

Place the pineapple quarters on 4 plates. Refrigerate.

Combine the chilled cream and vanilla sugar in a deep bowl. Set it in a larger bowl filled with ice water. Whip approximately 5 minutes with an electric beater until the cream forms stiff peaks.

To make the tree trunk, add the whipped cream to a pastry bag fitted with a large nozzle. Remove the plates from the refrigerator. On each one, make a strip about 2 inches wide and 4 inches long, parallel to the pineapple. Dust sifted cocoa over the cream strips.

Carefully stick the pineapple leaves in the cream, making the shape of a pine tree. With a smaller nozzle, dot the leaves with cream to create snowballs. Scatter on colored sprinkles.

Preparation: variable
No cooking necessary

1 fresh pineapple
³/₄ cup heavy cream, chilled
2 teaspoons vanilla sugar
  (see page 7)
Cocoa powder
Colored sprinkles

SWEET AND SAVORY CONDIMENTS

Chefs from all over the world have been using fruit in savory dishes for centuries. The history of French cuisine includes sweet-sour and salty-sweet flavor combinations such as duck à l'orange, duck with peaches or cherries, pork with prunes, and game with jellied currants. Fruit condiments have also been adopted by the English, who readily serve applesauce, cranberry sauce, or chutneys with their meat dishes. Americans, for example, like to have ham with pineapple. In the Antilles, bananas and papayas are used in savory and spicy dishes. In my travels across the world, I have never stopped discovering these enchanting combinations.

When I started my career in Paris at the Plaza-Athénée, I was captivated by a marvelous ham prepared in the American style: cooked bone-in and then covered in pineapple slices, which had been studded with cloves and dusted with brown sugar. The ham was roasted in the oven and came out deeply glazed, the pineapple transferring its taste to the ham while helping to preserve the ham's tenderness.

In Jamaica, we used a lot of fruit in our cooking, including plantains, those large, less sweet bananas. In Africa, in Kenya, Rhodesia, and other former British colonies, I had Indian assistants. I always saw them sweetening the spicy seasoning of their curries with coconut milk, or a banana or papaya purée.

Curries are a good example of how fruit can be used to its full potential as part of a larger meal. In these strong dishes, fruit pieces and sauces lay side by side on and around the plate in the form of condiments or other accompaniments, enhancing each other without blending. Each plate is a study in contrasts: the fruit sweetens the pepper, the chutney's heat tempers the coolness of the raw cucumber, the sugar complements the salt, the crunchiness of raw ingredients brings out the tenderness of cooked ones, and so on.

CHUTNEYS

Condiments that combine fruit and vegetables in a rich sweet-sour blend, chutneys are used like a mustard. They accompany pork, beef, lamb, fowl, and also enhance curry dishes.

Nothing should stop you from combining many fruits: golden apples and pineapple, apricot and mangoes, plums and pears. Always proceed in the same way, with the same amount of sugar and vinegar. Placed in jars like jellies, chutneys will keep for 1 to 2 months in the refrigerator.

Peach-Papaya Chutney
(Chutney de Pêches et de Papaye)

3 yellow peaches
1 papaya, weighing as
   much as the 3 peaches
1 clove garlic
1 onion
2 tablespoons grated fresh
   ginger
1 tablespoon chopped green
   pepper
1/4–1/3 cup cider vinegar
Juice of 1 lime
1/3 cup brown sugar
1/2 tablespoon salt

Peel and pit the peaches. Slice into sections. Peel, seed, and cut the papaya into chunks. Peel and chop the garlic and onion.

In a saucepan over medium-high heat, combine all of the ingredients. Stir in 1/4 cup water. Bring to a boil. Reduce to very low heat and cook for 30 minutes, skimming the top regularly. The chutney is ready to be poured into jars when the mixture is thick and syrupy. Pour into hot jars and seal.

Apricot Chutney

(Chutney d'Abricot)

Pit the apricots and slice into sections. Peel and cut the onion into thin strips. Seed the pepper and dice. Lightly crush the peppercorns.

Combine all the ingredients in a medium saucepan. Bring to a boil. Reduce heat to medium and cook for 25–30 minutes. When done, the chutney should have the consistency of jam. Pour into hot jars and seal.

2¼ pounds apricots
1 onion
1 yellow pepper
3 tablespoons grated fresh ginger
½ whole nutmeg, ground
½ teaspoon pink and green peppercorns
2 cups white wine vinegar
3 cups brown sugar
1 teaspoon powdered cinnamon

Sweet and Sour Morello Cherries

(Griottes en Aigre-Doux)

Store these morello cherries for at least 3 months at room temperature before eating them. Then, do not wait too long—they will dry out rather quickly. Use them like a chutney; they go very well with game. In Provence, we like to eat them with aperitifs.

Cut the cherry stems halfway. Wash, dry, and place the fruit in jars.

Bring the vinegar to a boil. Stir in the sugar and bay leaves. Pour over the cherries while boiling. Seal the jars airtight.

4½ pounds morello (sour) cherries
4¼ cups red wine vinegar
2¾ cups sugar
3 bay leaves

Equipment
Mason jars with rubber closings

Fruit Vinaigrette for Shellfish

(Vinaigrette aux Fruits pour Crustacés)

1/2 papaya, peeled and seeded
1/4 mango, peeled and pitted
1/4 melon, seeded and rind
    removed
1/2 peach, peeled and pitted
2 tablespoons balsamic
    vinegar
2 tablespoons red wine
    vinegar
3/4 cup olive oil
1 tablespoon Dijon mustard
Salt, to taste
Freshly ground pepper,
    to taste

Here's a delicious, quick, and unusual way to add flavor and refreshment to shellfish, whether warm or cold. Try it with grilled or steamed lobster, crab, scallops, or shrimp.

Cut the papaya, mango, melon, and peach into chunks. Blend all the ingredients together in a food processor until smooth. Season with salt and pepper to taste.

Pork Marinade

(Barbouille pour Travers de Porc)

3 cloves garlic
7 ounces apricot jam
3/4 cup ketchup
3 1/2 tablespoons
    Worcestershire sauce
2 tablespoons grated fresh
    ginger or 1 teaspoon
    powdered ginger

Great around the barbecue, this zesty marinade will also add flavor when grilling game.

Peel and finely chop the garlic. Combine all the ingredients in an airtight jar and refrigerate.

    A half hour before grilling, brush both sides of the meat with the marinade and set on a rack before cooking on the grill or under the broiler.

Fruit Mustard
(Moutarde de Fruits)

This delicious recipe came from my friends Nadia and Antonio Santini, owners of the restaurant Dal Pescatore situated in the tiny Italian village of Canneto (barely 70 inhabitants), which borders Oglio, halfway between Cremona and Mantua. They were willing to give me their version of this traditional condiment. In Italy, it is used as an accompaniment to stewed meats. To make it, you need mustard extract, available at gourmet and specialty food stores. Allow the mustard to mellow 5 to 10 days before eating.

2$^{1}/_{4}$ pounds fruit, slightly underripe (such as pears and apples)
2 cups sugar
9 to 12 drops mustard extract

Wash and peel the fruit. Cut into slices.

Combine the fruit and sugar in a bowl. Set aside for 8 hours.

Pour whatever juices have been released into a saucepan and bring to a boil. Immediately pour back over the fruit.

Set aside another 8 hours. Repeat the same steps 3 times over the next 24 hours.

Pour the syrup into a saucepan over medium heat. Gradually add the fruit, stirring, until it caramelizes. Transfer to a large container and let cool.

Add drops of the liquid mustard extract, to taste.

Store in an airtight jar in the refrigerator.

Strong Mustard with Orange
(Moutarde Forte à l'Orange)

6 sweet oranges
$^1/_2$ cup brown sugar
$2^1/_4$ pounds strong mustard

This tangy mustard will perk up many dishes: roasted fowl (especially goose and duck), pork, and grilled fish. It keeps 2–3 weeks in the refrigerator. Vary the recipe: substitute apricots cooked in honey syrup or peaches cooked like the orange rinds. Choose fruit that is not too ripe and still a bit firm. Make sure the fruit mixture is completely cooled before combining with the mustard.

Peel the oranges, including the pith. Place the rinds into a saucepan of cold water. Bring to a boil. Drain in a colander and run under cold water to cool. Dry on a towel. Cut into dice and set aside.

Juice the oranges (remove the seeds). Pour the juice into a saucepan with the sugar and the diced orange rind. Cook over medium heat until reduced to a thick syrup. Let cool thoroughly. Mix into the mustard. Store in jars.

A Few Ideas for Using Fruit in Savory Dishes

At the Moulin I serve a lobster *pastilla* with spinach leaves and dried apricots in a light turmeric cream sauce. It is a perfect blend of sweet and spice. I also use dried apricots to stuff a layered terrine of foie gras.

With a pork, rabbit, or duck terrine, use fruit for a sweet-and-sour taste.

Green, unripe mangoes cut into thin slices will bring an aromatic sour note to fish salads, shellfish, or poultry.

Prunes go especially well with bay leaves and pepper, and also complement the taste and bouquet of meats cooked in red wine. For example, add prunes to a baby rabbit stew cooked in a full-bodied wine 20 minutes before the rabbit is done. Do the same with pork, duck, or beef stew.

Blueberry jam is a good condiment for pork, turkey, and other roasted fowl, and especially game.

Applesauce is simply an apple compote (made with Granny Smith apples or russets) that is lightly salted, not sweetened, and topped with a dab of chilled butter. The British serve it with roast pork, goose, or duck.

A French tradition is to serve peeled blood sausage (boudin) fried in butter with melting, slightly browned apples (mostly pippins). I use the same amount of apple as sausage.

Because of its musky taste, duck is especially good with peaches, cherries, and oranges.

Fruits soften strong cheeses. Enjoy thinly sliced pear with Roquefort, grapes with hard cheeses (such as Comté, Beaufort, and Cantal), black cherry jam with baby goat cheese from the Pyrenees (an old Basque tradition!), and ripe figs with Parmesan.

I also love eating a hazelnut with a good cup of coffee.

CONVERSION TABLES

Liquid Equivalents

American Spoons and Cups	Liquid Ounces	Liquid Grams
1 tsp (teaspoon)	$^1/_6$	5
1 Tb (tablespoon)	$^1/_2$	15
1 cup (16 Tb)	8	227
2 cups (1 pint)	16 (1 pound)	454
4 cups (1 quart)	32	907
6 $^2/_3$ Tb	3 $^1/_2$	100
1 cup plus 1 Tb	8 $^1/_2$	250
4 $^1/_3$ cups	2.2 pounds	1000 (1 kilogram)

Flour Weights: Approximate Equivalents (scoop-and-level method)

3 $^1/_2$ cups of flour	1 pound	454 grams
1 cup	5 ounces	140 grams
$^3/_4$ cup	3 $^1/_2$ ounces	105 grams
$^2/_3$ cup	3 $^1/_4$ ounces	90 grams
$^1/_2$ cup	2 $^1/_2$ ounces	70 grams
$^1/_3$ cup	1 $^1/_2$ ounces	50 grams
1 Tb	$^1/_4$ ounce	8 $^3/_4$ grams
3 $^3/_4$ cups	17 $^1/_2$ ounces	500 grams or $^1/_2$ kilo

Sugar Equivalents

1 Tb	12 grams	½ cup	100 grams
¼ cup	50 grams	1 cup	200 grams
⅓ cup	65 grams	2 ¼ cups	454 grams (1 pound)

Butter Equivalents

1 Tb	½ ounce	⅛ stick	15 grams
2 Tb	1 ounce	¼ stick	30 grams
4 Tb	2 ounces	½ stick (¼ cup)	60 grams
8 Tb	4 ounces (¼ pound)	1 stick (½ cup)	113 grams
16 Tb	8 ounces (½ pound)	2 sticks (1 cup)	217 grams
	16 ounces (1 pound)	4 sticks (2 cups)	

Cup–Deciliter Equivalents: 1 deciliter equals 6 ⅔ tablespoons

Cups	Deciliters	Cups	Deciliters
¼	0.56	1 ¼	2.83
⅓	0.75	1 ⅓	3.0
½	1.13	1 ½	3.4
⅔	1.5	1 ⅔	3.75
¾	1.68	1 ¾	4.0
1	2.27	2	4.5

INDEX

Page numbers in *italics* refer
to illustrations.

A

almonds, *52*, 166
apples, 24–25, *68*
  Apple Compote, 64, *65*
  Apple Fritters, 118, *119*
  Apple-Quince Compote, 64
  Apple Tarts with Extra-Thin
    Crust, 86, *87*
  My Mother's Apple Rice
    Pudding, 121
  Pippin Apple Crumb Tart, *90*, 91
  Russet Charlotte with Corinth
    Grapes, 111
  Strawberry-Apple Fruit Drink,
    132
apricots, 12
  Apricot Chutney, 183
  Apricot Compote with Four
    Flavors, 58
  Apricot Fritters, 120
  Apricot-Lime Fruit Drink, 131
  Apricot Marmalade with Their
    Sweet Almonds and Vanilla,
    156
  Apricot Mousse, 106
  Apricot-Peach Jam with
    Vanilla, 152
  Apricots and Cream, 176
  Apricot Sauce, 137
  Apricot Sorbet, 141
  Apricots with Vanilla or
    Verbena, 173
  Apricot Tart, *78*, 79
  Sautéed Apricots, 98, *99*
  Sweet Apricot Clafoutis with
    Almond Milk, *94*, 96

B

bananas, 32
  Banana Fritters, *117*, 120
  César's Banana Salad with
    Wine and Honey, *50*, 51
  Pineapple-Banana Fruit Drink,
    132
blueberries, 20

C

cakes
  Bigarreau Cherry Cake with
    Black Pepper Cream, 102

  Pineapple Upside-Down Cake,
    100, *101*
  Rolled Cake, 177
candy, 165–71
  almonds, 166
  candied cherries, 165
  Crystallized Currants, 170
  Crystallized Flowers and
    Leaves, 170
  Crystallized Pansies, *171*
  dates, 166
  Fruit Jelly Candy, *168*, 169
  Kumquats in Multicolored
    Chocolate, 167, *168*
  prunes, 166
  small bitter oranges, 166, *166*
  strawberries, 166
  walnuts, 165
cherries, 9, 11
  Bigarreau Cherry Cake with
    Black Pepper Cream, 102
  candied, 165
  Cherries in Brandy, 173
  Cherry Compote with Kirsch
    and Almonds, 58
  Cherry Soup from Gordes, 55
  Moulin Cocktail, 132
  My Wild Black Cherry
    Clafoutis, 97
  Sweet and Sour Morello
    Cherries, 183
children, in the kitchen, 175–79
chocolate
  Chocolate-Coated Pear and
    Vanilla Ice Cream, 176
  Kumquats in Multicolored
    Chocolate, 167, *168*
  Pear Tart with Chocolate
    Crust, 84–85
  Pear Tart with Cocoa, 84
chutneys. *See* condiments
citrus fruit, 28–31, *29*, *30*, *60*
  Apricot-Lime Fruit Drink, 131
  Bitter Orange Jelly, 164
  Blood Orange and Currant
    Aspic with Sauternes Wine, 53
  Grapefruit Granita with
    Vermouth, *140*, 142, *143*
  Lemon Jelly with Verbena
    Leaves, 164
  Lemon Raspberry Tart, 76, *77*
  Lemon Sorbet, 138

  Orange Terrine, 114, *115*
  peeling, 40
  small bitter oranges, candied,
    166, *166*
  Strong Mustard with Orange, 186
clafoutis
  My Wild Black Cherry
    Clafoutis, 97
  Sweet Apricot Clafoutis with
    Almond Milk, *94*, 96
compotes, 54–65
  Apple Compote, 64, *65*
  Apple-Quince Compote, 64
  Apricot Compote with Four
    Flavors, 58
  Cherry Compote with Kirsch
    and Almonds, 58
  Dried Fruit Compote, 59
  Fig Compote, 61
  Mirabelle Plum Compote with
    Violets, 63
  Peach Compote with Three
    Flavors, 61
  Prune Compote with Wine and
    Tea, *62*, 63
  Quince Compote, 56
  Rhubarb Compote with Rum
    and Ginger, 56, *57*
  Strawberry Compote with
    Orange Blossom, 60
condiments, 181–87
  Apricot Chutney, 183
  Fruit Mustard, 185
  Fruit Vinaigrette for Shellfish,
    184
  Peach-Papaya Chutney, 182
  Pork Marinade, 184
  Strong Mustard with Orange,
    186
  Sweet and Sour Morello
    Cherries, 183
conserves, 172–73
  Apricots with Vanilla or
    Verbena, 173
  Cherries in Brandy, 173
  Mirabelle or Reine Claude
    Plums in Brandy, 173
  Purple Figs with Bay Leaf, 172
coulis, 46, *117*, 128–29
crusts, pie. *See* pastry, for tarts
currants, 15, 17
  Black Currant Crumb Tart, 91
  black currants, 17
  Blood Orange and Currant
    Aspic with Sauternes Wine, 53
  Crystallized Currants, 170